D1257167

# Life Cycles of Coccidia
of Domestic Animals

# Life Cycles of Coccidia of Domestic Animals

By Yevgeniy M. Kheysin

Edited by Kenneth S. Todd, Jr.
Translated by Frederick K. Plous, Jr.

University Park Press

BALTIMORE • LONDON • TOKYO

Library of Congress Catalog Card No. 71-156229
ISBN 0-8391-0066-3
Copyright ⓒ 1972
by University Park Press
All Rights Reserved
Printed in the United States of America

*Library of Congress Cataloging in Publication Data*

Kheĭsin, Evgeniĭ Mineevich.
    Life cycles of coccidia of domestic animals.
    Translation of Zhiznennye t͡sikly kokt͡sidiĭ domashnikh zhivotnykh.
    Bibliography: p.
    1. Coccidia. 2. Veterinary parasitology.
I. Title.
QL368.C7K4813      593'.19      71-156229
ISBN 0-8391-0066-3

# TABLE OF CONTENTS

# EDITOR'S PREFACE

The Coccidia have been the subject of many studies and reports, but no recent book is available that is concerned primarily with their basic biology and life cycles. *Zhiznennye Tsikly Koktsidii Domashnikh Zhivothykh* was published in Russian in 1967. The author, Dr. Yevgeniy Minevich Kheysin, was eminently qualified to write such a book. Much of the literature which he discussed was published in Russian and was assimilated with the non-Russian literature for the first time. Dr. Kheysin spelled his name Cheissin when he published in non-Russian journals; his name is listed as Kheisin in the *Index Catalogue of Medical and Veterinary Zoology*.

About one year after the book was published, Dr. Kheysin died of a heart attack in Leningrad. At the time of his death he was the Director of the Microscopy Laboratory of the Institute of Cytology of the USSR Academy of Sciences at Leningrad and Professor of Invertebrate Zoology at Leningrad University.

The following information was taken from a biography honoring Professor Kheysin on his 60th birthday by Yu. I. Polyanskiy (Tsitologiya 9:1436–1437, 1967). The biography was translated by Virginia Ivens.

Dr. Kheysin graduated from Leningrad University in 1930, and was a student of Valentin A. Dogel'. His first publications were concerned with the taxonomy and morphology of astomatid ciliates from oligochaetes. Some of his earlier investigations were on the parasitic amoebae of man. He was also interested in the piroplasmids and ciliates.

He is probably best known for his work with coccidia. His doctoral dissertation described the life cycles and cytology of several species of coccidia, and he was the first Soviet scientist to use the electron microscope to study protozoa.

His research was not limited to protozoa. He studied the permeability of parasitic helminth eggs, and he and a number of students and co-workers investigated the distribution and ecology of ixodid ticks in the territory of Karel.

Dr. Kheysin trained many students, and for several years he taught several zoology courses at the Pedagogical Institute in Leningrad. He was head of the Department of Zoology at Petrozabodshi University for six years.

He was well known outside of the USSR, and he and Yu. I. Polyan-skiy revised Dogel's books on general parasitology and general proto-zoology. Both of these were later translated into English. He partici-pated in the 14th Zoological Congress in London and the 15th Zoologi-cal Congress in Washington. He also took part in the International Protozoology Conferences in Prague and London, and in other interna-tional conferences. At the time of his death he was the vice-president of the Third International Congress of Protozoology which was held in Leningrad in August, 1969.

The coccidia have been of interest since the first species was described in 1870. They are important disease agents in many domestic animals, and large amounts of time and money have been spent to find methods for controlling them. The best method still seems to be good sanitation. An understanding of the fundamental ecology of this group is needed, and it is hoped that this book will provide much information and make it unnecessary to read the voluminous literature on the coccidia, much of which is published in languages other than English.

Mr. Plous translated the book and I have edited it. The opinions are those of Dr. Kheysin and not of Mr. Plous or me. The transliteration system for Russian names and words used is that of the United States Board on Geographic Names. The bibliography includes the references listed by Dr. Kheysin; however, I have corrected some of the citations by referring to the original articles that are available in the University of Illinois Library. The rest were verified by using the *Index Catalogue of Medical and Veterinary Zoology.*

The bibliography contains many references that were not cited in the Russian text. There are errors in some of the citations; these were in the original, and I did not attempt to verify that the correct papers were cited by Dr. Kheysin. The Russian edition contained 18 halftone plates, but these are not included in this book because of the poor quality of plates that are reproduced from halftones. Several of the line drawings in Kheysin's book were adapted from figures by other authors. For some of these I have copied figures from the original publications, and they are not the same as those in the original Russian book.

The translation of this book was supported by Research Grant CC00037 from the Communicable Disease Center, United States Public Health Service, and preparation of the manuscript was supported in part by a grant from the Stauffer Chemical Company and by Grant LM00798 from the National Library of Medicine.

Publication of this book was made possible by the assistance of sev-eral people. John V. Ernst, Datus M. Hammond, and Norman D. Levine reviewed the manuscript. The manuscript was typed by Margaret McWhorter and others of the stenographic staff of the College of Veterinary Medicine, at the University of Illinois in Urbana. I am especially indebted to my colleague Virginia Ivens for translation and interpretation of obscure parts of the text. Jack E. Snow assisted in proofreading the manuscript and in verifying the bibliography.

KENNETH S. TODD, JR.
*College of Veterinary Medicine*
*Urbana, Illinois*

# AUTHOR'S PREFACE

Because the group includes many parasites which cause serious diseases in man and domestic animals, representatives of the subclass Coccidiomorpha have long attracted the attention of researchers. It is sufficient to mention malaria and coccidiosis in order to emphasize the practical importance of this group of protozoa.

Many studies have been devoted to coccidiosis of domestic and wild animals, and many hundreds of different coccidian species have been described from vertebrates and invertebrates. During the last three decades many protozoology and parasitology texts have described coccidia and coccidiosis of domestic animals. In these texts considerable attention is given to the structure of the exogenous stages of development, pathogenesis, the clinical picture, treatment, immunity, prophylaxis, and the epizootiology and biology of the agents of coccidiosis.

In the Russian literature one should point out W. L. Yakimoff's (1931) book *Veterinarnaya Protozoologiya (Veterinary Protozoology)* which contained a large section on coccidiosis of domestic animals. However, much of the information given in this book is outdated. N. P. Orlov (1956) published a brochure on coccidiosis of farm animals. Also, a textbook on veterinary parasitology, compiled under the editorial leadership of Professor V. S. Yershov, contained a small section on coccidiosis.

In countries other than Russia, all the textbooks on parasitology and protozoology contain some information on coccidia and coccidiosis. Among these are books by Marotel (1949), Lapage (1956), and Borchert (1958) in which the coccidia are covered briefly along with other animal diseases. Coccidia and coccidiosis are treated in more detail in the specialized manuals of veterinary protozoology by Curasson (1943), Morgan and Hawkins (1948), Richardson and Kendall (1957), and Levine (1961a).

During the last 30 years, several monographs devoted exclusively to coccidia and coccidiosis have appeared. These include *Coccidia and Coccidiosis of Domesticated Game and Laboratory Animals and of Man* by Becker (1934), as well as the recently issued books *Coccidiosis* by Davies, Joyner, and Kendall (1963) and *Coccidia and Coccidiosis* by Pellérdy (1965). A monograph by Kheysin (1947a) dealt with the coccidia of rabbits, and Musayev and Veysov (1965) published a book on coccidia of rodents. In several major texts on protozoology (Wenyon, 1926; Doflein and Reichenow, 1953; Bhatia, 1938; Grassé, 1953; Jirovec et al., 1953; Kudo, 1954; Grell, 1956; Dogel', Polyansky, and Kheysin, 1962), the biological and cytological aspects of coccidia are examined.

It would seem that the coccidia have been examined from all aspects in many texts and special monographs, but information concerning the life cycles, principles of development within the host, and behavior in the external environment has not been sufficiently summarized. Most of the practical questions cannot be solved without a knowledge of the life cycles of coccidia, their morphophysiological properties, and the relationships between the various stages and the external environment. For this reason I have attempted to summarize the material now published and also to make use of my own data and experience with rabbit coccidia life cycles. Since coccidia are parasitic unicellular organisms, the greater part of the review of their life cycles is devoted to a cytophysiological analysis of all stages of the life cycles of these protozoa. Along with this analysis, and on the basis of it, a parasitological characterization of the life cycles is given.

In the following review, I shall describe not only the structure and physiology of all stages of the life cycle, but also their area of location in the body of the host, the characteristics of development and duration of the coccidian infection, the conditions necessary for the survival of the oocysts in the external environment and the effects of external factors on sporulation, and finally, the conditions determining infection of the host by coccidia.

In reviewing all these subjects, I have confined myself somewhat in the selection of material by chiefly using information on the coccidia of domestic animals. This is quite natural, considering that the life cycles of coccidia of domestic animals have been studied in relatively great detail, and because domestic animals are more convenient for experimentation. I have also used data on coccidia of invertebrate animals which are of a particular interest for purposes of comparison, since these protozoa offer the best way of studying certain cytological details of the life cycle such as the sexual process, reduction division, sporogony, and schizogony.

The last chapter of the book contains a description of all coccidia of domestic animals. The practical questions on methods of controlling coccidiosis are not included in the present work, but the reader may easily find them in the many books mentioned above.

# CHAPTER I

# The Basic Types of Life Cycles of Representatives of the Class Sporozoa

First of all, it is necessary to define the characteristics of the group of parasitic protozoa to which the coccidia belong. The term "coccidia" is so undefined that one must stipulate which organisms will be referred to by that name in this book.

The class Sporozoa includes the subclass Coccidiomorpha, or as it is otherwise called, the Coccidia. Thus, all representatives of this subclass may properly be called coccidia. However, in practice I have limited the use of this term somewhat. Sometimes all representatives of the order Coccidiida are referred to as coccidia. This is also proper, but with such a wide use of the term a certain misinterpretation and inconvenience arises because the order Coccidiida also includes the suborder which contains the blood sporozoa (Haemosporidiidea) which are not usually called coccidia. They are always referred to as haemosporidia. These include representatives of the genera *Plasmodium, Haemoproteus, Leucocytozoon,* and others. The term coccidia is better applied to the more limited group of sporozoans included in the order Coccidiida.

Very often only representatives of the suborder Eimeriidea are referred to as coccidia. This suborder includes numerous representatives which cause disease in various animals and man. None of these are parasites of the blood of the host, a fact which distinguishes them from the parasites of the suborder Haemosporidiidea, which are very similar to the Eimeriidea.

In the medical and veterinary literature the term coccidia is sometimes applied only to representatives of the family Eimeriidae, which includes all coccidia causing disease in man and animals. This family includes several genera and about 800 species. The disease caused by members of this family is called coccidiosis, but it would probably be more correctly called eimeriosis. Everyone has become accustomed to the former term, and practically no one uses the latter even though it is completely correct.

Further descriptions of coccidia will be limited to representatives of the family Eimeriidae which are parasitic in domestic animals. This includes a rather large group of parasitic protozoa on which a special monograph could be written.

The class Sporozoa is strictly parasitic and includes many intracellular forms which are found in both vertebrates and invertebrates. At the present, there are about 1800 known species of this class. The following taxonomic scheme will be used in this book:

**Class Sporozoa Leuckart, 1879**
    Subclass Gregarinina Dufour, 1828
      Order Eugregarinida Léger, 1900
      Order Schizogregarinida Léger, 1910
    Subclass Coccidiomorpha Doflein, 1901
      Order Protococcidiida Cheissin, 1956
      Order Adeleida Léger, 1911
         Family Adeleidae Léger, 1911
         Family Haemogregarinidae Wenyon, 1926
      Order Coccidiida Labbé, 1899
        Suborder Eimeriidea Léger, 1911
         Family Selenococcidiidae Poche, 1913
         Family Eimeriidae Poche, 1913
         Family Lankesterellidae Reichenow, 1921
         Family Aggregatidae Reichenow, 1921
        Suborder Haemosporidiidea Danilevskij, 1885
         Family Haemoproteidae Doflein, 1916
         Family Plasmodiidae Mesnil, 1908

The most characteristic feature of the Sporozoa is the presence of a complex cycle of development which includes an alternation of a sexual process with various forms of asexual reproduction. In all the Sporozoa, the sexual process is completed after fertilization and formation of an encysted zygote called a zygocyst or oocyst. In the Sporozoa this stage of the life cycle is the most resistant one and also the longest lasting.

The sexual process is characterized by the formation of gametes of each sex which may be structurally similar or different. In the Gregarinina, gametogenesis results in the formation of similar gametes; dissimilar gametes are formed in the Coccidiomorpha. In the Gregarinina the male and female gametes are formed by division of the gamont. In the Coccidiomorpha this process is found during microgametogenesis, but the macrogamete is formed by the growth of the gamont without division. When gamonts of both sexes are formed in the Gregarinina, or during microgametogenesis of the Coccidiomorpha, the gametocyte is the stage of development which produces gametes. In macrogametogenesis this term is not applicable. In the gregarines, the gamont stage (gametocyte) is present for some time before the process of gametogenesis begins. This stage of the gamont is often called the trophont or trophozoite.

The gametes formed by all the Sporozoa unite to form a zygote. The isogametes, i.e., gametes which cannot be distinguished in size or shape as being of either sex, unite. In anisogamic fertilization the gametes are morphologically distinguishable, but there is little difference in size; such gametes also unite. In oogamic fertilization there is a considerable difference in size of the gametes. The microgamete penetrates into the macrogamete, and a zygote is formed by the fertilized macrogamete. Isogamic, and in some cases anisogamic, fertilization is usually found in gregarines, but oogamic fertilization is characteristic of the Coccidiomorpha.

In all Sporozoa the zygote undergoes metagamic development, or sporogony, leading to the formation of sporozoites within oocysts (zygocysts). The process of sporogony takes place by either simple or multiple division. The sporozoites are the stage of development which infect the host and represent the initial stage of the next generation which reproduces by either an asexual or sexual process. Asexual reproduction is accomplished by schizogony (multiple division). The sporozoite penetrates a cell of a host tissue to become a uninuclear trophozoite. During the process of asexual reproduction (schizogony), the single nucleus divides repeatedly. This leads to the formation of a multinuclear schizont. Segmentation of the latter then begins, i.e., the breakdown of all the schizont protoplasm into separate formations called merozoites (the French protistologists call them schizozoites) which are equal in number to the number of nuclei formed. The merozoites are usually mononuclear. They may penetrate into the cells of the host where they form schizonts. This type of repeated asexual reproduction produces a large number of parasites within the host. Asexual reproduction by schizogony sooner or later changes into a sexual process which begins the progamic period. The merozoites then develop into gamonts.

Thus, the developmental cycle of the Sporozoa may include a subsequent alternation of gametogony, sporogony, and repeated asexual reproduction (agamogony) by schizogony with the formation of merozoites. This type of developmental cycle occurs in most of the Coccidiomorpha (except for the order Protococcidiida) and in the Schizogregarinida. In addition to this type of complete sexual cycle, some Coccidiomorpha, e.g., Protococcidiida, and all the Eugregarinida may have an abbreviated cycle characterized by an absence of asexual reproduction. In such a life cycle, gamogony and sporogony are present and take place in the same manner as in the complete cycle of development. However, when the sporozoites enter the tissue of the host they do not develop into schizonts but become gamonts.

An oocyst is always produced in both the complete and abbreviated cycles, and nuclear reduction division (meiosis) takes place during the first metagamic division of the zygote. As a result, all stages of the developmental cycle of Sporozoa are haploid, except for the zygotes, the nuclei of which are diploid.

The Sporozoa have the ability to form spores during the metagamic period. The name "Sporozoa," suggested by Leuckart (1879–1886) was based on the fact that at a certain period of development a spore stage appears in which the sporozoites develop and become capable of infecting a new host. The spores are those stages of development which have a protective covering. Leuckart named those stages of Sporozoa which have protective coverings "spores." Even today gregarine oocysts are erroneously referred to as spores. The spores of bacteria, fungi, spore-bearing plants, Cnidosporidia, and Sporozoa are different structures and are not homologous. The spores of bacteria, fungi, mosses, Cnidosporidia, and Sporozoa can be grouped together only in the sense that all of them have protective coverings. This is a convergent feature, developing independently in various groups of animals and plants and at different stages of growth. [Although Kheysin used the words "spores" and "sporocysts" synonymously, sporocysts will be used in the rest of the book. Ed.]

Among the Sporozoa, the protective coverings are first found on the oocysts which are eliminated into the external environment. The walls of the oocysts serve to protect the zygote against the effects of the environment. The protective walls of the oocyst are formed around the zygote or even around the macrogamete prior to fertilization.

The process of sporogony, resulting in the formation of sporozoites, takes place within the oocyst. In all the gregarines eight sporozoites, lying directly under the wall of the oocyst, are formed. The same is found in several representatives of the Coccidiomorpha. In the oocysts of *Schellackia, Lankesterella, Cryptosporidium, Tyzzeria, Dobellia,* and in all the Haemosporidiidea, the sporozoites form directly under the oocyst wall.

In many members of the Coccidiomorpha, a sporocyst is formed within the oocyst, and sporozoites are formed in the sporocyst. The sporozoites are surrounded by a solid wall which forms on the surface of the sporocyst. The presence of sporocysts inside the oocysts provides the sporozoites with an additional protection against the effects of unfavorable external factors. Oocysts and sporocysts are formed in the genera *Mantonella, Caryospora, Cyclospora, Isospora, Eimeria, Wenyonella, Dorisiella, Hoarella,* and other members of the family Eimeriidae. Therefore, the presence of a sporocyst cannot be made the basis for characterizing the class Sporozoa, since more than half of the representatives have no sporocysts.

Some stages of Sporozoa have the ability to move. The intracellular stages of development can move by amoeboid locomotion. The merozoites of various Sporozoa move by stretching and contracting the body; this is brought about by the contractile elements in their cytoplasm. Trophozoites and gamonts of the gregarines are able to creep, an ability also associated with the presence of contractile elements in the surface layers of the body. Flagellar locomotion is observed only in the male gametes of certain coccidia. The gametes of many gregarines display amoeboid movement.

The Sporozoa obtain nourishment through all the body surfaces. There are still no reliable data to indicate the existence of special organelles in the cytoplasm. It is possible that the micropores observed with electron microscopy in the sporozoites of *Plasmodium, Lankesterella,* in the merozoites of *Eimeria intestinalis,* and in the trophozoites of *Eucoccidium durchoni* (Protococcidiida) are microcytostomes (Kheysin, 1965), but this hypothesis needs verification (Garnham, Baker, and Bird, 1962; Vivier and Henneré, 1965). The schizonts of *Plasmodium* are capable of phagotrophy (Rudzinska and Trager, 1957). Apparently, they ingest the stroma of erythrocytes by pinocytosis.[1] The possibility cannot be eliminated that other Sporozoa may also have the same capability.

The Sporozoa represent a group of Protista well delimited from all the other Protozoa. The presence of a sexual process, oocysts, and spo-

---

[1] Supplement to correction. When the manuscript was set in type, Aikawa et al. (1966) reported on the mechanism of nutrition of schizonts and gamonts of *Plasmodium*. Electron microscopy revealed the presence of cytostomes through which the cytoplasm of host cells is ingested.

rogony is perhaps most characteristic of these parasitic organisms which have a complex cycle of development. All Sporozoa can be separated into two natural groups representing subclasses, both of which have a common origin and a close physiological similarity. The difference between the two subclasses is in their method of gametogenesis. In the subclass Gregarinina the process of gametogenesis is the same when gametes of either sex are formed. Gametes of both sexes, regardless of whether isogamy or anisogamy takes place, are always formed in equal numbers. There may be one such gamete for each sex, e.g., *Ophryocystis,* or a greater number. When many gametes are formed, those of both sexes arise from the corresponding gametocytes by multiple division. The nuclei are distributed at the periphery of the gametocyte, and with the cytoplasm form motile gametes. A large part of the gametocyte cytoplasm is lost at this point in the form of a residual body which is not used in the formation of the gametes.

Regardless of the degree of sexual differentiation of gametes, after fertilization a zygote (covered with a protective wall) is formed; this develops into an oocyst. During the process of sporogony, sporozoites are formed in each oocyst. The oocysts containing the sporozoites serve to infect a new host. In the intestine of the host, the sporozoites excyst from the oocyst and penetrate into epithelial cells of the intestine where initial development takes place. There they become trophonts. Intracellular existence changes into extracellular existence [sic]. In the Eugregarinida the trophonts are gamonts. In the Schizogregarinida the sporozoite develops into a schizont, and the merozoites, which are a result of schizogony, develop into gamonts.

The gamonts of the gregarines are usually grouped in pairs in syzygy and are surrounded by a protective wall to form gamontocysts. Within the wall, gametes are formed from the gametocytes inside, and after fertilization zygotes and oocysts (zygocysts) are formed. All gregarines parasitize invertebrate animals. The general scheme of the life cycle of the Sporozoa is unique in the gregarines (Fig. 1).

The other subclass, Coccidiomorpha, which includes a large number of individuals parasitic in both invertebrates and vertebrates, differs from the gregarines in that during gametogenesis unequal numbers of male and female gametes are always formed. In addition, oogamy takes place, i.e., the female gametes have the form of macrogametes and the male gametes assume the form of microgametes. Only one macrogamete is formed from each female gamont and during this process no progamic division occurs. In all representatives of the Coccidiomorpha, the female gamete, because of the accumulation of a large amount of

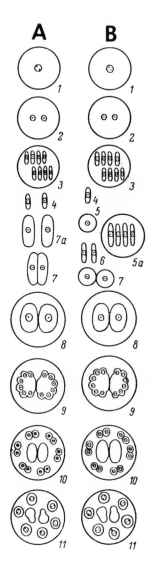

Fig. 1. Developmental cycle of sporozoans of the subclass Gregarinina. *A*, Eugregarinida; *B*, Schizogregarinida. *1*, Oocyst; *2*, first metagamic division; *3*, sporulated oocyst; *4*, sporozoites; *5*, trophozoite; *5a*, mature schizont; *6*, merozoites; *7a*, gametocytes; *7*, gametocytes combined into syzygies; *8*, encysted syzygy (gamontocyst); *9*, gametogenesis; *10*, union of gametes; *11*, zygote.

reserve food material, reaches a relatively large size as compared with the very small and motile flagellated microgametes. The latter, unlike the macrogametes, are always formed from the gamont by multiple division in a number exceeding that of the macrogametes. In some of the Adeleida, progamic division may lead to the formation of two or even four microgametes. In *Klossiella cobayae,* two microgametes are formed, but in some species of *Adelina* there are four. Among representatives of the suborder Haemosporidiida, 12 microgametes are formed. In the suborder Eimeriidea, as well as in the order Protococcidiida, several dozen and even hundreds of microgametes are formed from one gamont during microgametogenesis. Because only one macrogamete is formed from one gamont, it is not necessary to refer to the gamont as a gametocyte. Here the concepts of macrogamont and macrogamete are similar. In microgametogenesis the quantitative and qualitative difference between microgametocytes and microgametes is distinct.

Fertilization of the Coccidiomorpha is accomplished by the microgamete penetrating into the macrogamete. The zygote which results, as in gregarines, becomes encysted and develops into an oocyst. During sporogony, from four to several hundred sporozoites are formed. After penetration into the host the sporozoite may develop either into a gamont, as in Protococcidia, or into a schizont in all the other coccidia. The developmental cycle of the Coccidiomorpha is schematically presented in Fig. 2.

Thus, the most stable part of the sporozoan life cycle is the sexual process and sporogony. Sporogony is similar in all of the Sporozoa with only insignificant deviations. The sporozoites may be formed not directly in the oocyst, but in sporocysts, the number of which may vary in different genera as may the number of sporozoites.

The asexual part of the cycle may be absent in both the gregarines (e.g., Eugregarinida) and Protococcidiida. Since asexual reproduction of the Sporozoa leads to an increase in their productivity, it is natural that a breakdown of this process of the life cycle is possible only in the presence of some kind of compensatory mechanism which provides great numbers of infective stages. In the Eugregarinida maximal fecundity is accomplished by uniting the gamonts of both sexes in syzygy within one wall. Under these conditions it is possible for the gametes to fertilize, even though they are relatively immobile and there are not greater numbers of male gametes. Also, the large amounts of nutritive materials accumulated in the gametocytes permit the formation of large numbers of gametes. The absence of asexual reproduction in gregarines is compensated for by a considerable amount of sexual reproduction, and by

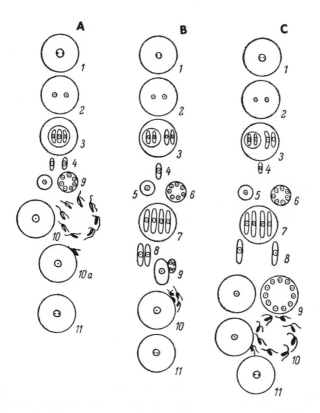

FIG. 2. Developmental cycle of sporozoans of the subclass Coccidiomorpha. *A*, Order Protococcidiida; *B*, order Adeleida; *C*, order Coccidiida. *1*, Oocyst; *2*, first metagamic division (meiosis); *3*, sporulated oocyst; *4*, sporozoites; *5*, trophozoites; *6*, polynuclear schizonts; *7*, segmented schizont; *8*, merozoites; *9*, macrogamete and polynuclear microgametocyte; *10*, macrogamete and microgamete; *10a*, fertilization; *11*, zygote.

the fact that a very large number of gametes is formed from one female gametocyte.

On the other hand, in the Schizogregarinida one female gamont forms a relatively small number of gametes and a correspondingly small number of oocysts (from 1 to 14 in the majority; only in *Syncystis, Lipocystis,* and *Sawayella* can 128, 256, and up to 8000 oocysts be formed). Schizogony increases the productivity of these gregarines by creating a possibility for a larger number of gametocytes.

Among the coccidia, asexual reproduction is absent only in the Protococcidiida (Fig. 2, *A*). This is compensated for by the high productivity during the metagamic period. For example, in the oocysts of *Eucoccidium dinophili,* as many as 250 sporocysts, each with six sporozoites, are formed. When ingested by the worm which is the host of this parasite, one oocyst brings in about 1500 sporozoites. When one considers the small size of the host, such infections are sufficient for repetition of the developmental cycle and maintenance of the existence of the species.

In most of the coccidia (orders Coccidiida and Adeleida) the life cycle consists of asexual reproduction, which provides for a large number of macrogametes which develop from the merozoites of one or more generations. The order Protococcidiida is characterized by the absence of asexual reproduction. The sporozoites develop directly into gamonts which develop extracellularly in the body cavity or intestine. The rest of the development of Protococcidiida is typical of the subclass Coccidiomorpha. Representatives of the order are found in annelids and other invertebrates. The type species is *Eucoccidium dinophili* Grell, 1956. Development of this species takes place in *Dinophilus gyrociliatus.* The sporocysts release sporozoites in the intestine; these penetrate into the body cavity, where they develop into macro- and microgamonts. The macrogamont enlarges and becomes a macrogamete. The microgametocyte also grows, and 32 microgametes form at its surface. After fertilization, an oocyst, which contains 250 sporocysts, forms (Fig. 3). In addition to this species, *Eucoccidium* (syn., *Coelotropha*) *durchoni* Vivier, 1963, has been described from the body cavity of *Nereis diversicolor* and *Eucoccidium ophryotrochae* Grell, 1960 from *Ophryotrocha puerilis.* The developmental cycle of these species is very similar to that of *E. dinophili.*

Apparently, several other coccidia, e.g., *Angeiocystis, Merocystis, Myriospora, Caryotropha,* and *Pseudoklossia* (for which only a sexual process and sporogony have been described) may also be assigned to the Protococcidiida if further research shows that they lack schizogony. The same is true of *Coelotropha vivieri* Henneré, 1963 from the coelom of the polychaete *Notomastus latericeuns.* This coccidium is very similar to *Eucoccidium* and only the presence of a small suction rostrum on the trophozoites serves as the basis for assigning it to a separate genus.

Apparently, this order may also include the single representative of the genus *Eleutheroschizon (E. duboscqui)* which is parasitic in the intestine of polychaetes. Schizogony has not been observed in these parasites. The development of macro- and microgamonts takes place on the surface of the intestinal epithelium. The mononuclear, oval macrogametes are ready for fertilization after passing from the digestive tract of the worm into the surrounding water. The multinuclear microgametocytes

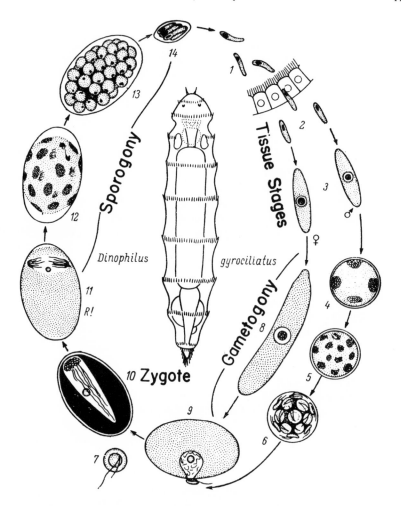

FIG. 3. Developmental cycle of *Eucoccidium dinophili* (After Grell, 1956). *1*, Sporozoites in the intestinal lumen; *2*, sporozoites in the body cavity; *3*, macrogametes and microgametocyte; *4* and *5*, polynuclear microgametocytes; *6*, mature microgametocyte; *7*, microgamete; *8*, large macrogamete; *9*, fertilization; *10*, zygote in prophase of the first metagamic division; *11*, reduction division; *12*, oocyst; *13*, oocyst with many sporocysts; *14*, sporocyst with sporozoites.

are formed in the intestine of the polychaete, and then passed into the external environment, where a large number of biflagellated microgametes are formed. Fertilization and oocyst formation also occur in the water. During sporogony, a large number of sporozoites are formed in the oocyst, but no sporocysts are formed. The properties of the sexual process in *Eleutheroschizon* are such that no doubt remains as to the affiliation of this genus with the subclass Coccidiomorpha (Grassé, 1953), and not with Gregarinia, as was suggested by Reichenow (Doflein and Reichenow, 1953).

Schizogony is always part of the life cycle of representatives of the orders Adeleida and Coccidiida and the subclass Coccidiomorpha. The sporozoites develop into schizonts and not into gamonts as in the Protococcidiida, and the development of asexual generations and gamonts usually takes place within the cells. There are considerable differences between the orders Adeleida and Coccidiida.

It is characteristic of Adeleida that the macro- and microgamonts develop in direct contact with each other. A young microgamont combines with a growing macrogamete and both gamonts are surrounded by a common shell similar to the syzygy of the gregarines. The microgamont forms two or four microgametes. The gametes formed in this manner are always capable of penetrating into the macrogamete, since they lie directly on its surface. Sporogony, like schizogony, proceeds according to the plan common to all sporozoans (Fig. 4).

Some representatives of the Adeleida (e.g., family Adeleidae) are parasitic in land and fresh-water invertebrates (insects, centipedes, mollusks, leeches, oligochaetes). The parasites develop in the cells of the intestine, the fat body, the Malpighian tubules, the musculature, the hypodermis, etc. The family Adeleidae includes the genus *Klossiella,* which is parasitic in vertebrate hosts. One species, *K. muris,* is a common parasite of mouse kidneys; another, *K. cobayae,* parasitizes the renal tubules of guinea pigs. Representatives of the family Haemogregarinidae are parasitic in vertebrates in both the endothelium of the blood vessels and in the parenchymal cells of the liver, as well as in the formed elements of the blood (primarily the erythrocytes). All haemogregarines develop in two hosts, and the vectors are invertebrate animals. The haemogregarines parasitic in the blood of aquatic reptiles, amphibians and fish are spread by leeches, and the haemogregarines of land reptiles are transferred by ticks. Schizogony takes place in the vertebrate host, where the gamonts are also formed, but the gametes are formed and fertilization takes place in the vector. The oocysts are formed and sporogony takes place in the same vector. Species of *Karyolysus* are spread by gamasid mites. In the vertebrate host, schizogony takes place in endothelial cells, but the gamonts are found in erythrocytes. The sexual cycle and sporogony also occur in mites.

Several species of the genus *Hepatozoon* are parasitic in liver, spleen,

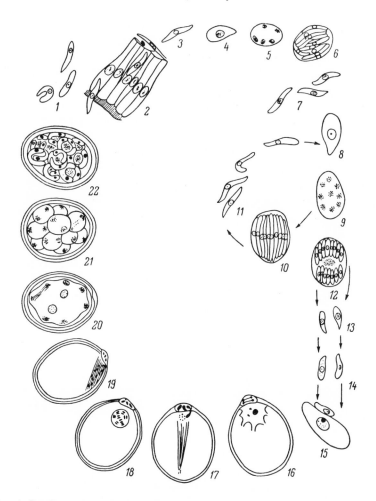

Fig. 4. Developmental cycle of *Adelina cryptocerci* (After Yarwood, 1937). *1,* Sporozoites; *2,* sporozoites in the epithelium; *3* to *6,* schizogony; *7,* merozoites; *8* to *11,* new generation of schizogony; *12,* mature schizont with telomerozoites; *13,* gamonts; *14,* macrogametes and microgametocytes; *15,* union of macrogamete and microgamete; *16,* formation of four microgametes; *17,* fertilization; *18* to *19,* zygote and reduction division; *20* to *22,* sporogony.

and bone marrow cells of mammals (e.g., rats, dogs), or in some reptiles, where schizogony occurs. They are spread by gamasid mites and ixodid ticks, as well as by several mosquitoes and flies. The gamonts develop in erythrocytes and leukocytes of the vertebrate host, but the sexual cycle occurs in ticks or other vectors. The oocysts formed in the body of the tick increase in size during sporogony and form a large number of sporocysts with numerous sporozoites.

In a third order, the Coccidiida, the macro- and microgamonts develop independently and separately from one another. The microgametes are always formed in greater numbers than in the Adeleida. This is especially pronounced in those coccidia in which microgametogenesis occurs within cells (Figs. 5 to 9). Under these conditions there is the greatest differentiation of macro- and microgametes. Usually a multitude of motile, flagellated gametes are formed, and gamete fertilization is made possible. If the gametes are formed outside the cells, such as in the cavity of the intestine of arthropods, as occurs with the blood Sporozoa, the number of microgametes is usually not large (from 6 to 12). However, under such conditions, even this small number is sufficient to fertilize the macrogametes.

The order Coccidia includes the blood Sporozoa (suborder Haemosporidiidea), which have an alternation of hosts, and the suborder Eimeriidea of which the majority develop in one host only. The suborder Haemosporidiidea includes the genera *Haemoproteus*,[2] *Leucocytozoon,* and *Plasmodium.* In these genera, the gametes always form in an invertebrate host (an insect), but outside the cells in the lumen of the digestive tract (in the middle portion of the intestine). The microgametes are flagellum-shaped and form on the surface of the residual body. The zygote is always motile and might best be called a zygokinete (often it is erroneously referred to as an ookinete, but this term presupposes the motility of the egg, or macrogamete, and in actuality it is the zygote that moves). The oocysts develop within the body of the insect, and during the process of sporogony they increase in size considerably. Several thousand sporozoites are produced. In the vertebrate host the asexual generations develop both within and outside of erythrocytes.

Members of the suborder Eimeriidea are chiefly parasites of vertebrates, but there are a small number of species which are also parasitic in invertebrate hosts. Unlike the Haemosporidia, almost all the eimeriids develop without a change of hosts. However, it is characteristic of some representatives that they have two invertebrate hosts *(Aggregata)* or a vertebrate and an invertebrate host *(Schellackia, Lankesterella).* In the latter case, development of the parasite does not occur in the invertebrate host (gamasid mites or leeches). Sporozoites are present in the intestine of the invertebrate, but further development is possible only if

---

[2]The genera *Hepatocystis, Nycteria,* and *Polychromophilus* apparently should be included in the one genus *Haemoproteus* (Bray, 1957).

they return to the vertebrate host. Usually the developmental stages of the Eimeriidea are located within cells, most often in the mucosa of the intestine (less often in several other internal organs, e.g., liver, kidneys, etc.). In *Selenococcidium* the growing schizonts and early gamonts are found outside cells. In *Cryptosporidium* all the stages develop within the cells; the same is true of *Eimeria gadi* and *Eimeria pigra*.

In most of the Eimeriidea the oocysts pass into the external environment. This is the only stage of development which is found outside the host. All the other developmental stages, such as the schizonts and gamonts, develop within the host. In all the Eimeriidea the motile microgametes form in large numbers on the surface of the microgametocyte. In the overwhelming majority of cases the zygote is immobile, and the oocyst, with its solid shell, does not grow in size during sporogony (sometimes the oocyst may swell slightly).

Representatives of the suborder Eimeriidea may be divided into several families characterized by several specific properties of their life cycles. In the family Selenococcidiidae development occurs in one host, and the growing schizonts and gamonts are found outside the cells. One species, *Selenococcidium intermedium* (Fig. 5), is parasitic in the intestine of lobsters. Vermiform schizonts develop in the lumen of the intestine. After the formation of eight nuclei, they penetrate the cells of the intestine where they become spherical and develop into a small number of vermiform merozoites. The latter again enter the lumen of the intestine where they increase in number by an asexual process. After two or three asexual generations, two sizes of schizonts are formed. The smaller ones form eight small merozoites which enter the lumen of the intestine and penetrate into the epithelial cells to form microgametocytes. A large number of biflagellated microgametes form on the microgametocyte surface. The large schizonts produce four vermiform merozoites. The latter penetrate into intestinal cells and become macrogametes. After fertilization, a round oocyst is formed in which the sporozoites develop. This process, however, has been poorly studied. Infection of lobsters occurs when mature oocysts are eaten.

The family Eimeriidae is characterized by the development of the parasite in one host and by passage of the oocysts into the external environment. Usually the schizonts and gamonts are intracellular and do not have a vermiform shape. This family includes at least 800 species which are parasitic in both vertebrate and invertebrate hosts. The general scheme of this family's developmental cycle (Fig. 6) is as follows: the sporozoite penetrates into the epithelial cells of the intestinal mucosa or other organs, where it develops into a schizont. The schizont grows, and

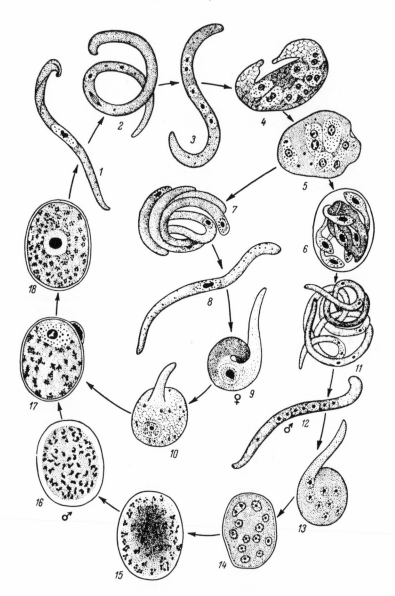

FIG. 5. Developmental cycle of *Selenococcidium intermedium* (After Léger and Duboscq, 1910). *1*, Mononuclear schizont; *2* to *5*, growing schizont (penetrating into intestinal cells); *6* and *11*, mature schizont with merozoites, forming what will eventually be microgamonts; *7*, merozoites forming macrogametes; *8* to *10*, development of macrogametes; *12* to *16*, microgametogenesis; *17*, fertilization; *18*, formation of oocysts.

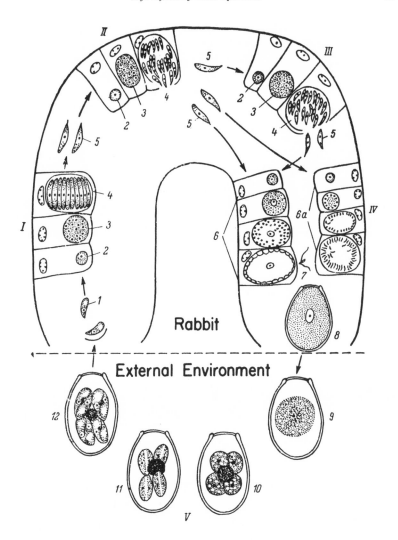

FIG. 6. Developmental cycle of *Eimeria magna* (After Kheysin, 1963). *1*, Sporozoite; *2* to *4*, first and succeeding generations of schizonts; *5*, various generations of merozoites; *6*, development of macrogamete; *6a*, microgametogenesis; *7*, microgamete; *8*, zygote (oocyst); *9*, oocyst in external environment; *10* to *12*, sporogony. I to III, Schizogonic reproduction; IV, gametogony; V, sporogony.

its nucleus divides. It then forms merozoites, the number of which corresponds to the number of nuclei in the adult schizont. At the very least, two to four merozoites may be formed; the maximum is around 100,000.

The merozoites emerge from the cell in which they were formed and lead an extracellular life for a short time. They then penetrate into different epithelial cells of the same organ. After they enter cells, asexual reproduction (schizogony) starts over, and this process may be repeated several times. Finally, the merozoites penetrate into the cells of the host and develop not into schizonts, but into gamonts. Some of these become macrogamonts (macrogametes); the rest become microgametocytes and form a multitude of microgametes at their surfaces. Fertilization takes place within the cells, after which a protective covering forms around the macrogamete which develops into an oocyst. In most cases the oocysts pass into the external environment where sporogony takes place. In some Eimeriidae the oocysts sporulate in the host. In either case infection is brought about when the host ingests sporulated oocysts which are present in the external environment. The structure of the oocysts is quite varied both as to the number of sporocysts and number of sporozoites. Among representatives of this family are a small number of species which are pathogenic for domestic animals and man.

The family Lankesterellidae includes approximately 15 species. All of these develop with an alternation of vertebrate and invertebrate hosts. The oocysts do not pass into the external environment, but the sporozoites penetrate into erythrocytes which may be ingested by a vector. Gamasid mites serve as the vectors for the genus *Schellackia,* and gamasid mites and leeches are the vectors for the genus *Lankesterella.* Sporozoite development does not occur within the vector. The sporozoites accumulate in the epithelial cells of the intestine and enter a vertebrate host when the vector is eaten.

The developmental cycle of *Schellackia* is similar to that of *Eimeria.* Schizogony and gametogony occur in the epithelium of the intestine (Fig. 7). However, the macrogametes penetrate into the connective tissue, where they become oocysts after fertilization. It is in the oocyst that sporogony takes place, and the sporozoites thus formed penetrate into the capillaries of the cardiovascular system and are distributed throughout the body. When gamasid mites ingest blood, the sporozoites enter their intestines where they are retained but do not multiply. The vertebrate hosts of *Schellackia* are several varieties of lizards which become infected by eating the mites.

Representatives of the genus *Lankesterella* are parasitic in amphibians and birds (Lainson, 1959, 1960). The schizonts develop in the endothelial cells and macrophages of the spleen, liver, bone marrow, and other organs. The macrogametes and microgametocytes develop in these same cells. They are especially numerous in the cells of the reticuloen-

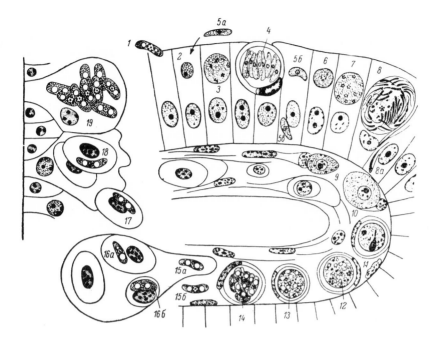

FIG. 7. Developmental cycle of *Schellackia* in a lizard and the gamazid mite, *Liponyssus saurarum* (After Reichenow, 1921a). *1*, Sporozoite penetrating into an epithelial cell; *2 to 4*, schizogony; *5a to 5c*, merozoites; *6 to 8*, microgametogenesis; *8a*, microgamete; *9*, macrogamete; *10*, fertilization; *11*, first metagamic division of the zygote nucleus; *12 to 14*, sporogony; *15a* and *15b*, sporozoite penetrating into the bloodstream; *16a* and *16b*, sporozoite in an erythrocyte; *17* and *18*, erythrocyte with sporozoites phagocytized by epithelial cells of mite intestine; *19*, several sporozoites in epithelial cell of the mite intestine.

dothelial system of the liver, kidneys, and lungs. After fertilization, a zygote, covered with a delicate shell, develops in these same cells. A large number of sporozoites is formed within the oocyst. They emerge from the oocysts into the capillaries and penetrate monocytes and lymphocytes circulating in the blood. The sporozoites enter a vector, where they develop (Fig. 8). Leeches are the vectors for *Lankesterella* species which are parasitic in amphibians, and gamasid mites are the vectors for those species of *Lankesterella* of which birds are the host.

In the family Aggregatidae, development occurs in two invertebrate hosts. The oocysts are passed into the external environment. In the intestine of a crab the sporozoites pass through the epithelial cells and into the connective tissue, where they become schizonts (Fig. 9). Here

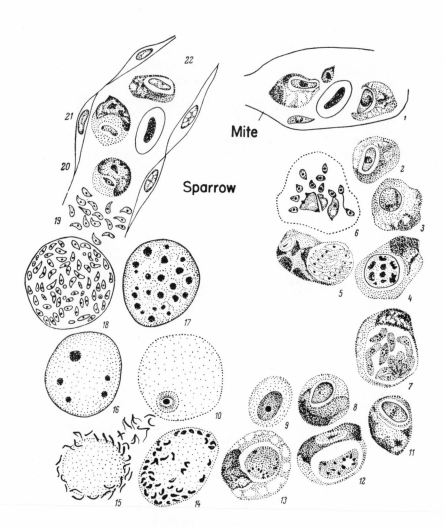

FIG. 8. Developmental cycle of *Lankesterella garnhami* (After Lainson, 1959). *1*, Sporozoites in ingested leukocytes and free in the intestine of a gamazid mite, *Dermanyssus gallinae; 2* to *6*, schizogony in macrophages of the spleen, liver and bone marrow of a sparrow; *7*, schizogony yielding the large merozoites which form gamonts; *8* to *10*, development of macrogametes in macrophages; *11* to *15*, microgametogenesis in the same cells of liver, lungs, and kidney; *16* and *17*, division of zygote nucleus; *18*, mature oocyst; *19*, sporozoites in the liver, kidney, and lungs; *20* to *22*, penetration of sporozoites into monocytes and lymphocytes in the blood.

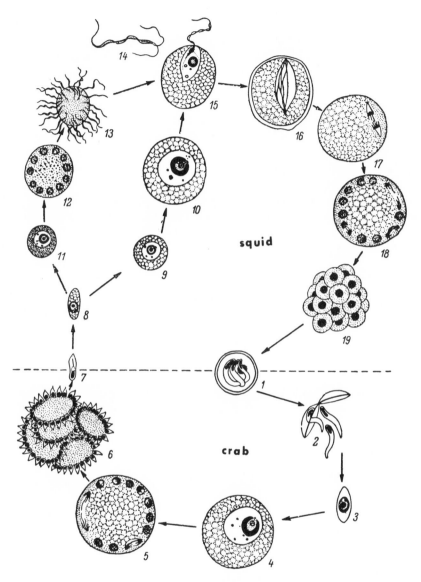

squid

crab

Fig. 9. Developmental cycle of *Aggregata* (After Dobell, 1925). *1*, Sporocyst with three sporozoites; *2*, excystation in the intestine of a crab; *3*, young schizont; *4* to *6*, schizogony; *7*, merozoite within a crab being ingested by a cephalopod mollusk; *8*, trophozoite in the intestinal wall cells; *9* and *10*, macrogamete growth; *11* and *13*, microgametogenesis; *14*, microgamete; *15*, fertilization; *16*, encystation of the zygote with diploid chromosome selection (2*n* = 12); *17*, first metagamic division (meiosis), formation of haploid nucleus; *18* and *19*, sporogony.

schizogony takes place, resulting in the formation of merozoites. When a cuttlefish eats the crab the merozoites penetrate through the epithelial cells of its intestine and into the connective tissue, where gamont development also takes place. Only the very earliest stages of the gamonts are present in the epithelium. The oocysts develop in the wall of the cuttlefish's intestine. Many sporocysts, each containing three sporozoites, form within each oocyst. The mature oocysts pass into the water and are eaten by crabs. At the present, about 20 species of this family have been described.

Among the representatives of the suborder Eimeriidea are some parasites of invertebrate animals. They include, for example, *Angeiocystis, Myriospora, Merocystis, Caryotropha,* and *Ovivora.* Only the sexual stages have been described for these coccidia, and schizogony is either unknown or may not be present (Vivier and Henneré, 1964b). It may be that they do not have a schizogonic cycle and develop exactly as does *Eucoccidium.* If so, they should be included in the order Protococcidiida. On the other hand, the opinion has been expressed that, as in the case of *Aggregata,* schizogony of these coccidia occurs in hosts which are still unknown (Wenyon, 1926; Bhatia, 1938). If this is true they should be included in the order Coccidiida (family Aggregatidae).

These four families of the suborder Eimeriidea will be objects for further study and description in this book.

The position of *Dobellia binucleata* Ikeda, 1914, which is parasitic in the intestinal epithelium of sipunculoids, is not entirely clear. Because the microgamont and macrogamete develop in contact with each other, Wenyon (1926) and Grassé (1953) included this genus in the order Adeleida. Doflein and Reichenow (1953) considered it a member of the Coccidiida, since the microgametocyte forms a large number of microgametes. Apparently, the latter authors are correct. The formation of a syzygy in *Dobellia* is a secondary phenomenon, but the large numbers of microgametes may be regarded as their primary feature. All the properties of *Dobellia* permit the species to be included in the family Eimeriidae.

# CHAPTER II

# Morphophysiological Properties of Various Stages of the Life Cycle of Coccidia of the Suborder Eimeriidea

Each stage in the developmental cycle of coccidia has certain specific properties of structure and metabolism. In addition, coccidia always have a strict and sequential progression of developmental stages accompanied by a change in the conditions of the habitat.

Within the host, the intracellular stages, such as the schizonts, always alternate with the extracellular merozoites which are formed from schizonts and begin their development within cells. The merozoites spend only a short time outside the cells in the lumen of the intestine or in the bile ducts. However, the intracellular development of the schizonts lasts for a considerable period. In some coccidia, e.g., *Selenococcidium,* the reverse is true, and the schizonts first develop in the intestinal lumen of the lobster and then enter cells. The latter phase is shorter in duration than the extracellular one. The schizonts and merozoites differ from each other greatly, not only in structure, but also in the nature of their metabolism. Great adjustments in structure and physiology occur in the transition from asexual reproduction to gamogony and later, after the formation of a zygote, to sporogony.

As a result of intracellular microgametogenesis, motile microgametes are formed; these leave the host cell and live outside the cells for a time. Their life span is very short and many microgametes perish without having penetrated a macrogamete; others participate in the process of fertilization. During the process of growth, the macrogamete is usually located within a cell. In rare cases it develops outside the cell, e.g., *Cryptosporidium parvum* develops outside the intestine of mice. An important part of macrogametogenesis is the preparation of nutritive and energy sources needed for subsequent existence of the encysted zygote which is formed by the macrogamete outside of the host organism after fertilization. The walls around the zygote isolate it from the surrounding environment and provide for further development either outside the cell or outside the host in the external environment. This takes place without any exogenous source of nutritional substances and under conditions differing markedly from those in which the macrogamete develops. Sporogony usually takes place under these conditions, and the stores of nutritive substances accumulated by the macrogamete are deposited in the sporozoites. These are needed by the sporozoites as a source of energy during subsequent extracellular existence when they leave the oocyst in the lumen of the intestine and penetrate into host cells. Also, a supply of energy is required for the process of excystation.

An interesting feature of the development of Coccidia and other Sporozoa is that each succeeding stage in the life cycle differs from the previous one by the presence of a series of new and specific organelles typical of only that stage, and by the disappearance or alteration of several structures which existed earlier. For example, the conoid, paired organelle and micropore of the merozoites, which are easily seen only with the electron microscope, apparently disappear in the schizonts and gamonts which develop from the merozoites. These organelles develop during schizogony, but how this takes place and from what structures they arise are still unknown. The microgametocyte, which develops from the merozoite, does not have the fibrillar elements which are characteristic of merozoites. During the course of microgametogenesis, the microgametes form basal bodies and flagella which are absent in the previous stages of development. It is likewise impossible to observe any structures in the macrogamete which were present in the merozoites. Thus, in the entire cycle of coccidial development, there is a complex metamorphosis during which morphogenesis takes place, and the end of the cycle is eventually completed under the harsh conditions of the external environment.

## ASEXUAL REPRODUCTION

### Schizonts and Schizogony

Asexual reproduction of coccidia is by schizogony, which is a process of multiple division. It begins with a mononuclear stage of a young schizont (trophozoite). Growth then begins, and during this time nuclear division takes place. The numerous nuclei of the schizont scatter throughout the schizont cytoplasm in a random matter, but in some cases they move to the periphery. This multinuclear stage is the adult or mature schizont. Later, a portion of cytoplasm forms around each individual nucleus and the entire schizont more or less synchronously segments or breaks into merozoites, the number of which is equal to the number of nuclei formed as a result of multiple division. During schizogony, in those cases in which the nuclei are situated at the periphery, the merozoites seem to bud off from the surface of the schizont, and a residual body is formed, e.g., *Eimeria hirsuta* from the intestine of the larva of the *Gyrinus* beetle. A residual body also forms when the nuclei do not concentrate on the surface. In such a case, the residual body is smaller.

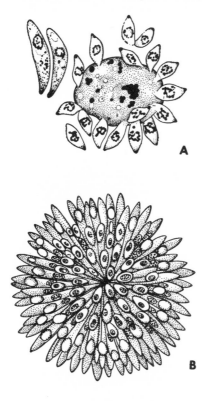

Fig. 10. Segmented schizonts of various species of coccidia. *A*, Second-generation schizont of *E. praecox* (After Tyzzer et al., 1932); *B*, schizont of *E. hirsuta* with merozoites located around the residual body (From Grassé, 1953).

The mature schizonts of *E. intestinalis* and *E. magna* both have a small residual body, whereas those of *E. praecox* (Tyzzer et al., 1932) and *E. hirsuta* (Fig. 10) or *E. gilruthi* (Chatton, 1910) have a large residual body.

Most mature schizonts are oval or spherical. Schizonts developing in host cells are isolated from the host cell cytoplasm by a delicate membrane which is visible only with the electron microscope. Segmentation of the schizont results in the formation of merozoites within this thin wall. The merozoites inside the schizont may move actively and, apparently by their movements, may tear open the wall. As a result, the merozoites emerge from the schizont, and an empty space or vacuole, shaped like the former schizont, remains in the host cell.

The sizes of schizonts differ to a large degree among various species and even between different generations of one species. Small schizonts, up to 10 microns in diameter, are found in *E. perforans, E. irresidua, E. anguillae, Cryptosporidium parvum,* and *C. muris.* The smallest schizonts (5 to 8 microns) are found, as a rule, in those species of coccidia in which development occurs on the surface of the epithelium in the area of the striated border, e.g., *Cryptosporidium, E. anguillae,* and others.

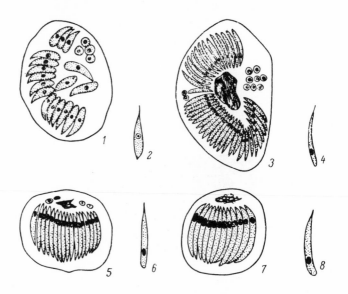

FIG. 11. Schizonts and merozoites of different generations of *E. intestinalis* (After Kheysin, 1948). *1* and *2,* First-generation schizonts and merozoites; *3* and *4,* second-generation; *5* to *8,* third-generation schizonts and merozoites.

Schizonts having a diameter of 20 to 30 microns are the most common (Figs. 11, 13, 14, and 54). If schizonts develop in the epithelium, their size is limited to a certain degree by the size of the epithelial cells. In some cases, the epithelial cells containing schizonts become deformed and increase in size to the point that the schizont in such a cell appears larger than the schizonts from unchanged cells. For example, second generation schizonts of *E. intestinalis* are larger (15 to 35 microns), than those of the first generation (13 to 15 microns) which do not change the size of the infected cell. Likewise, in *E. irresidua* the schizonts are 12 to 13 microns in diameter but do not cause any changes in the epithelial cells.

The largest schizonts (over 100 microns in diameter) develop in the subepithelium of the intestinal mucosa. Such giant schizonts, which are visible to the naked eye, are found in *E. cameli* from camels, *E. bovis* from cattle, *E. gilruthi* from sheep (Fig. 12), *E. leuckarti* from horses, *E. travassosi* from *Dasypus sexcinctus, E. navillei* from *Tropidonotus [Natrix] natrix,* and *E. christenseni* from goats. These species were earlier combined into one subgenus, *Globidium* Flesch, 1883, only because they have huge schizonts which were given the name globidial schizonts. The dimensions of such schizonts reach 300 by 170 microns in *E. leuckarti,* 280 by 150 microns in *E. christenseni,* 0.5 mm in *E. gilruthi,* 100 by 84 microns in *E. travassosi,* 0.3 to 0.4 mm in *E. bovis,* and 78–250 by 70–150 microns in *E. auburnensis.* They resemble a cyst within a solid shell. Large numbers of merozoites form inside such schizonts. According to Hammond et al. (1946), an average of 120,000 merozoites develops in first-generation schizonts of *E. bovis.*

Fig. 12. Small part of a "globidial" schizont of *E. gilruthi* (From Grassé, 1953).

The formation of merozoites in the large schizonts of *E. gilruthi* takes place around small cytoplasmic centers (blastophores or cytomeres). After the merozoites are formed the blastophores develop into residual bodies which sometimes contain nuclei which did not form merozoites. A similar breakdown of the schizont into cytomeres or meroblasts is observed in *Caryotropha mesnili* in the body cavity of the marine worm, *Polymnia nebulosa,* as well as in *Aggregata eberthi* in the intestine of the crab. However, Vivier and Henneré (1964a) were of the opinion that *Caryotropha* actually had schizonts. They hypothesized that the so-called schizonts were actually oocysts with numerous sporocysts and sporozoites.

The number of merozoites into which the schizont breaks up depends to a large degree on the schizont size. The giant schizonts form thousands of merozoites, while the smaller ones form dozens or at the most several hundred. Such comparisons should be made only within one species, since it may be that the same size schizont in different species may have varying numbers of merozoites. This property is genetically predetermined for each species, as is the size of the merozoites formed in a single schizont.

It is possible to present a series of examples of the relationship between the number of merozoites and the size of schizonts. In the large second- and third-generation schizonts of *E. intestinalis,* which have a diameter of 15 to 25 microns on the 7th day after inoculation, 70 to 120 delicate merozoites which are 6 to 10 microns long and 0.5 to 0.8 microns wide develop. At the same time, the intestine still contains other schizonts, 13 to 16 microns in diameter with 30 to 70 merozoites that are 10 to 12 microns long and 1.5 microns wide (Figs. 11 and 14).

In *E. magna,* fourth-generation schizonts are 25 to 40 microns in diameter and form 120 merozoites which are 6.3 microns long and 2 microns wide, and second-generation schizonts are 14 to 20 microns in diameter and form 6 to 40 merozoites that are 8.9 microns long and 2.1 microns wide. In *E. irresidua,* first-generation schizonts develop in the epithelium and have a diameter of 10 to 13 microns. They form 12 to 20 merozoites that are 6 to 10 microns long and 1.5 to 3 microns wide. Second-generation schizonts have a diameter of 6 to 10 microns and produce only 5 to 15 merozoites that are the same size as those of the first generation. However, if the schizont divides into small merozoites, their number is greater than that of large merozoites from schizonts of the same size. In *E. media,* first-generation schizonts have a diameter of 12 to 24 microns and form 6 to 11 large merozoites which measure 8 by 2 microns; second-generation schizonts are the same size and form 13 to

18 merozoites that are 7 microns long and 1.5 microns wide. The large third-generation schizonts of this species measure 18 to 25 microns and form 30 to 100 merozoites that are 6 to 7 microns long and 1.0 to 1.3 microns wide (Kheysin, 1947b).

First-generation schizonts of *E. adenoeides* from turkeys measure 17 by 13 microns and contain 80 to 100 merozoites which are 4.5 by 1.5 microns. Other schizonts, which appear within 48 hours after infection, measure 8 by 7 microns and break down into 8 to 16 merozoites that are 7 microns long and 1.5 microns wide (Fig. 56).

*Eimeria brunetti* will be used as an example of the large coccidia. One generation of schizonts measures 20 by 16 microns and forms 50 to 60 merozoites, while another generation has small schizonts that measure 10 by 9 microns and form only 12 merozoites (Boles and Becker, 1954). In *E. tenella* first-generation schizonts measure 24 by 17 microns and form as many as 900 merozoites which are 3 microns long. Second-generation schizonts, although larger (about 50 microns in diameter), form 200 to 300 merozoites that are up to 16 microns long (Fig. 52).

The schizonts of each species of coccidia have their own specific structural properties. They differ from one another in size, shape, and number of merozoites that are formed. To a lesser degree, it is also possible to find differences in the structure of their nuclei and cytoplasm.

Within each species there are differences between schizonts of different generations. It is also possible that differences exist in the structure of schizonts that form merozoites of different sexual potentials. It is quite possible that sexual differentiation, which is said to occur in the asexual generations, may express itself in the different structures of schizonts.

Schizonts of various generations have different potentials. The first generation always yields merozoites which develop into subsequent asexual generations, while the second and third generations, and in some cases even the fourth and fifth generations, form merozoites which develop into either gamonts or schizonts.

Nabih (1938) suggested a special terminology for describing the various generations of schizonts. The object of his research was *Klossia loosi,* a representative of the order Adeleida. He suggested calling the first-generation schizonts protoschizonts and their reproduction proto-schizogony. The following several asexual generations, which form the real schizonts, were given the name schizonts or euschizonts; the last generation, which gives rise to the gamonts, was called teloschizonts, and the process of their formation was called teloschizogony. It is difficult to apply this terminology to the coccidia of the suborder Eimeriidea, since it is not always possible to establish the number and order of the generations, and in addition the schizonts of one generation, as can be judged from the time of their formation, may sometimes form merozoites of varying potential. Some yield new generations of schizonts while others yield gamonts. Thus, it is better to speak of schizonts of the first, second, third, and succeeding generations.

The question as to which generation a schizont belongs can be determined by the time of its formation after infection and by its structural characteristics. However, in some cases where there is a lack of synchrony in development, and polymorphism of schizonts and merozoites is present, it is rather difficult to determine the number of asexual generations.

In first-generation schizonts which develop directly from sporozoites, it is sometimes possible to observe a solid protein body during the transition from sporozoites into schizonts. This has been observed in *E. tenella, E. meleagridis, E. meleagrimitis, E. media, E. nieschulzi,* and *E. separata* (Tyzzer, 1929; Kheysin, 1947b; Clarkson, 1959a and b; Roudabush, 1937). This body is located at one pole of the sporozoites and is somewhat broadened and rounded.

First-generation schizonts of *E. bovis* appear within 2 weeks after infection and are huge in size (281 by 203 microns). The second-generation schizonts are observed within another 1.5 to 2 days and are extremely small (8.9 to 10 microns) and form no more than 30 to 36 merozoites (Hammond et al., 1963).

In *E. tenella* first-generation schizonts appear in the ceca between 48 and 72 hours[1] and measure 24 by 17 microns. They break down into 900 merozoites. The second generation develops within 96 to 120 hours; the schizonts reach a diameter of 50 microns and contain 200 to 300 relatively large merozoites (Fig. 52). At a later time, it is possible to find small third-generation schizonts, with 4 to 30 merozoites measuring 7 microns; these are situated under the nuclei of the epithelial cells (Tyzzer, 1929). Some of the second-generation merozoites and all of the third-generation merozoites of *E. tenella* form gamonts. However, Gill and Ray (1957) reported that *E. tenella* does not have a third generation of schizonts and merozoites, but that the small schizonts thought to be the third generation result from sexual differentiation of the schizonts and are the origin of the microgametocytes.

In *E. brunetti* first-generation schizonts appear on the 2nd day. They

---

[1] Here and henceforth when hours or days are used, the appearance of a given stage of development is calculated from the time of ingestion.

measure 30 by 20 microns and form up to 200 merozoites. On the 4th day, second-generation schizonts, measuring 20 by 16 microns with 30 to 60 merozoites, are formed. On the 5th or 6th day a third generation develops, consisting of small schizonts that are 10 by 9 microns and contain 12 merozoites (Fig. 51).

In *E. necatrix* first-generation schizonts are found in the small intestine 2.5 to 3 days after infection, and on the 5th to 8th day large second-generation schizonts (52 by 38 microns) are present; these form both gamonts and schizonts. At this time mature third-generation schizonts with 6 to 16 merozoites are present. *Eimeria acervulina* and *E. maxima* yield two generations of schizonts.

*Eimeria meleagrimitis* from turkeys produces three generations of schizonts. The first generation is complete by the 48th to 66th hour. These schizonts measure 17 by 13 microns and break down into 80 to 100 merozoites that are 4.5 by 1.5 microns. On the 3rd day small second-generation schizonts measuring 8 by 7 microns with 8 to 16 merozoites appear. The third generation matures by the 96th hour. These schizonts are the same size as those of the second generation and form the same number of merozoites. The second-generation schizonts develop below the nuclei, but those of the third generation develop above the nuclei of the epithelial cells (Clarkson, 1959a).

*Eimeria adenoeides* and *E. meleagridis* produce only two generations of schizonts. By the 54th to 60th hour the first species has mature first-generation schizonts which are 30 by 18 microns and contain 700 merozoites. Between the 96th and 108th hour, mature second-generation schizonts appear; these are 10 by 10 microns and have 12 to 24 large merozoites (Clarkson, 1958, 1959b).

There are three asexual generations in *E. phasiani*. The first-generation schizonts mature within 48 hours after inoculation and form 50 to 100 merozoites. Within 66 hours, development of the second-generation schizonts is completed; these are 9.83 by 9.03 microns and contain 12 to 16 merozoites. The third generation matures by 84 to 90 hours. The schizonts have a diameter of 7.7 microns and produce 12 to 16 merozoites which have a diameter somewhat smaller than those of the second generation (5.9 by 1.5 microns and 7.43 by 1.44 microns, respectively) (Trigg, 1965).

Among rat coccidia there are three to four asexual generations (Roudabush, 1937). The first generation of *E. nieschulzi* develops within 36 hours and forms 20 to 36 merozoites; the second generation completes development within 48 hours and forms 10 to 14 merozoites; the third generation develops by the 72nd hour and yields 8 to 20 merozoites; and the fourth generation develops at the 96th hour and forms 36 to 60 merozoites. In *E. separata* first-generation schizonts mature within 24 hours. The second generation develops within 48 hours, and the third by the 77th hour. All the schizonts form 2 to 12 merozoites. The life cycle of *E. miyairii* is similar.

*Eimeria magna* in rabbits develops into first-generation schizonts by the 48th hour which measure 11 by 7 microns and yield 8 to 24 merozoites. Within 92 to 96 hours after infection, larger second-generation schizonts appear. They are 14 to 36 microns and form as many as 40 merozoites. At the 120th hour, even larger third-generation schizonts (25 to 40 microns) are present; these have 40 to 89 merozoites. In *E. intestinalis* the first-generation schizonts mature on the 4th to 6th day after infection; the schizonts are 15 to 25 microns and contain 20 to 60 merozoites. Beginning on the 5th day, the second-generation forms 90 to 120 merozoites. They are present until the 9th day. On the 8th to 9th day, small (13 to 16 microns) third-generation schizonts with 15 to 30 merozoites are found. In *E. piriformis,* mature first-generation schizonts appear in the large intestine of rabbits on the 6th day. They reach a diameter of 20 microns and break down into 15 to 25 merozoites. The second-generation schizonts are smaller (11 to 12 microns) and contain 25 to 55 merozoites; these are present on the 7th to 9th day after infection (Kheysin, 1947b, 1948).

Since there is no synchrony in the development of asexual generations, in some cases schizonts of different generations may be present in the cells of the intestinal mucosa at the same time. These schizonts may be structurally different. At the end of the prepatent period, various forms of schizonts may be found. This is determined not only by their affiliations with various generations, but apparently also by sexual dimorphism. No matter what generation the schizonts belong to, they are characterized by common properties of metabolism which depend upon the basic functions of growth and reproduction of the schizont. The young schizonts begin growing from a small, rounded body formed from sporozoites or merozoites. The mature schizont is many times larger than the young one. Consequently, during schizogony the mass of cytoplasm increases considerably. In addition to growth, there is an intense reproduction of nuclei by mitotic division. This process causes a need for deoxyribonucleic acid (DNA) at all stages of schizogony. Therefore, DNA synthesis occurs during schizogony. By the end of schizogony, all the complex structures of the future merozoites are formed. In connection with this, active synthetic processes are taking place during schizogony.

At all stages of growth, the schizont cytoplasm is highly basophilic; this is associated with the presence of ribonucleic acid (RNA) (Pattillo and Becker, 1955; Kheysin, 1958a, 1960). The high RNA content of the cytoplasm is due to increased protein synthesis during growth of the schizont and to the processes of differentiation which occur during formation of the merozoites. During segmentation of the schizont, RNA is distributed in the cytoplasm of the merozoites and is partially retained in the residual body.

The precursors of RNA (free purine ribonucleotides) are concentrated in the cytoplasm of the young schizonts. The amount of these precursors decreases as the schizonts grow. This was observed by Beyer (1963a) in second-generation schizonts of *E. intestinalis* and *E. magna* with the aid of Roskin and Balicheva's (1961) uranyl reaction. In addition to the cytoplasm, free purine ribonucleotides were observed in the karyosomes of nuclei, but the nucleoplasm gave only a weak positive reaction (Beyer, 1963a). In those cases where the karyosome of the nucleus was pronounced, it always contained a certain amount of RNA (Kheysin, 1958a, 1960).

The way in which coccidia synthesize free purine ribonucleotides is not known. It is possible that they may be synthesized from the host cell metabolites or even obtain it in prepared form. One indirect confirmation of the latter supposition might be the gradual decrease of free purine ribonucleotides in the epithelial cells of the rabbit's intestinal mucosa in which the schizonts of *E. intestinalis* develop (Beyer, 1963a). It is probable that the use of nucleic acids synthesized from precursors in the host cell is typical for many intracellular parasites (Moulder, 1962).

Cytoplasmic RNA plays a large role in the synthesis of protein, but it is probable that the RNA in the karyosome participates to a certain degree in this process. However, a karyosome has not been observed in the nuclei of schizonts of certain species of coccidia (e.g., *E. intestinalis*), and the synthesis of protein within these species apparently occurs at the same rate as in those schizonts which have karyosomes (e.g., *E. magna*).

In those parts of the cell in which intensive protein synthesis occurs, phosphatase is always present. In schizonts of *E. magna,* the nuclei produce a strong reaction to alkaline phosphatase, but this does not occur in the cytoplasm. In the nuclei, alkaline phosphatase activity is evident only in the karyosomes in which apparently an active protein synthesis occurs and in which a large part of the nuclear RNA is concentrated (Beyer, 1960). High alkaline phosphatase and RNA activity have also been observed in the karyosomes of the nuclei of *E. tenella* schizonts (Gill and Ray, 1954a; Ray and Gill, 1954; Tsunoda and Itikawa, 1955). Gill and Ray stated that this is an expression of the functional connection between alkaline phosphatase and RNA. In this case, protein synthesis, under conditions of high RNA content, is catalyzed by the enzymatic activity of phosphatase. It is weakly expressed in the nucleoplasm. On the other hand, the activity of acid phosphatase in the nucleoplasm of *E. intestinalis, E. tenella,* and *E. magna* schizonts is much higher than that of the alkaline phosphatase (Tsunoda and Itikawa, 1955; Beyer, 1960). In the karyosomes of *E. tenella* schizonts, the activity of acid phosphatase and 5-nucleotidase was relatively weak. In the cytoplasm of young *E. intestinalis* and *E. magna* schizonts, phosphatase was not present, but in polynuclear schizonts it was found not only in the nucleus but also in the cytoplasm (Beyer, 1960).

In order to synthesize material in the cytoplasm, it is necessary to have not only RNA but also proteins containing sulfhydryl groups. This has been found in metazoan cells and has also been observed in the cytoplasm of *Trypanosoma cruzi* (Kallinikova and Roskin, 1963). However, in coccidian schizonts, such parallelism in the distribution of proteins containing the SH-group and RNA has not been observed (Beyer, 1963). In schizonts, the RNA content does not decrease with growth, but the reaction to the SH- and SS-groups in the polynuclear schizonts decreases noticeably. It is highly probable that this occurs as a result of the intracellular environment of the schizonts. In merozoites which have left the host cell, the thiol compound content rises again. Since the presence of sulfhydryl groups in the cell is connected with the rate of the oxidation-reduction processes, it may be assumed that the content is not high.

In trophozoites and growing schizonts of *E. intestinalis, E. magna,* and *E. coecicola,* there is a negative reaction to succinate dehydrogenase activity (SDH) in both the nucleus and cytoplasm. In polynuclear schizonts of *E. intestinalis* and *E. coecicola,* the reaction to SDH is negative, but in *E. magna* there are small granules of monoformazan which increase in number and size as the schizonts mature (Beyer, 1963). The schizonts of coccidia lead a typical anaerobic existence; this can be demonstrated by their lack of succinate dehydrogenase activity. At the same time, the epithelial cells of the intestinal mucosa obtain energy by aerobic oxidation of the substrate. In these cells there is always a high SDH level (Beyer, 1963).

When the parasites are located within cells, the glycolytic respiration of nutritional substances moving from the cell into the cytoplasm of the parasite may completely satisfy the energy demands for synthetic processes. As the schizonts grow, the food materials of the host cell are exhausted and the parasite uses the remaining substrate by means of a mechanism which is energetically more efficient, i.e., it passes into aerobiosis. This happens whenever the schizont segments into merozoites. The SDH activity level increases in the cytoplasm of such intracellular merozoites. In *E. magna* the transition to aerobiosis is completed immediately after segmentation of the schizont, but in *E. intestinalis* and *E. coecicola* it occurs shortly before the merozoites emerge from the host cell into the lumen of the intestine (Beyer, 1963).

There is a relatively small amount of glycogen in the schizonts of *E. magna* and *E. intestinalis* which accumulates near the end of growth prior to segmentation of the schizont (Kheysin, 1958a, 1960). An analogy was observed in schizonts of *E. tenella* (Gill and Ray, 1954b; Rootes and Long, 1965). In some cases there is so little glycogen in the schizonts that it is difficult to find. Some investigators (Edgar, Herrick, and Fraser, 1944; Giovannola, 1934) failed to find any glycogen in the schizonts of *E. tenella, E. stiedae,* and *E. falciformis,* nor was any glycogen found in the schizonts of *E. brunetti* or *E. acervulina* by Pattillo and Becker (1955).

In the growing polynuclear schizonts of *E. necatrix,* the highest activity level of two enzymes playing an important role in carbon exchange was found. These were glucosan phosphorylase and glucosan transglycosylase (Rootes and Long, 1965). In mature schizonts which have accumulated glycogen, these enzymes have a lower level of activity. Apparently, the accumulation of glycogen in the schizont takes place at the expense of the glucose which is present in the host cells (Gill and Ray, 1954b; Rootes and Long, 1965). Lipids have not been observed in the cytoplasm of growing schizonts (Gill and Ray, 1954b; Pattillo and Becker, 1955; Kheysin, 1958a, 1960).

In schizonts of *E. magna,* the cytoplasm contains bacillus-shaped mitochondria (Kheysin 1940, 1947b). When stained with Janus green, the schizonts of *E. tenella* also show mitochondria (Nath, Dutta, and Sagar, 1965). Their ultrastructure is characterized by the presence of a comparatively small number of short tubules surrounded by two unit membranes. The tubules are outgrowths of the internal limiting membrane of the mitochondria (Kheysin, 1965; Scholtyseck, 1965b). The small number of tubules in the mitochondria is probably associated with the anaerobic existence of the schizonts (Kheysin, 1963).

In the various species of coccidia, the nuclei of schizonts are more or less similar in structure. In the interphase nucleus, the chromatin, which yields a positive Feulgen reaction, is situated at the periphery and forms one or two clusters. There is an easily seen Feulgen-negative karyosome in the center of the nucleus of some species which stains deep red with pyronin when a double strength of methyl green is used with Una's pyronin. The karyosome is especially large in the nuclei of *E. maxima* schizonts (Scholtyseck, 1963a). A karyosome is not visible in *E. intestinalis* schizonts.

The ultrastructure of schizonts has not yet been studied sufficiently. Scholtyseck's (1965b) data on the schizonts of *E. stiedae* and *E. perforans* are valuable. It is characteristic that the cytoplasm of both species has a large number of membranes in the endoplasmic reticulum which are situated concentrically around the nuclei. Apparently, the formation of fissures occurring along these membranes in the cytoplasm of the schizont leads to a separation of the merozoites during segmentation of the schizont. This process of merozoite formation by splintering the entire schizont into individual pieces is a characteristic feature of schizogony.

## Merozoites

Merozoites are formed by schizogony; these emerge from the intracellular schizont and lead an extracellular way of life for a short time. Because merozoites emerge from the schizont and abandon the host cell there is a possibility they may resettle within the same organ and penetrate into new host cells. Thus, the merozoites are a stage of development which aid in spreading the parasite within the host. In the intestinal coccidia, only the merozoites enter the intestinal lumen, where they can move significant distances from the place where they were formed.

Merozoites are a motile stage in the development of coccidia. The body shape of most merozoites is similar. Most often they are spindle-shaped *(E. intestinalis)*, cigar-shaped *(E. separata)*, sausage-shaped *(E. bovis)*, or banana-shaped *(E. meleagridis)*. Sometimes the small merozoites are shaped like a slightly stretched pear *(E. miyairii)* or a wide comma *(E. adenoeides)*. The anterior end of the body may be pointed, but the posterior end is rounded. Very often the cigar-shaped or spindle-shaped merozoites, especially if they are long, are slightly bent out of line like a sickle or boomerang.

The dimensions of merozoites vary within small limits. The smallest merozoites are 2 to 4 microns long (e.g., the first generation of *E. tenella*) while the largest ones reach 20 to 30 microns, (e.g., *E. neoleporis* Carvalho, 1944 and *Wenyonella africana* Hoare, 1933). The third-generation merozoites of *E. nieschulzi* (Fig. 13) are 17 to 21 microns long. The most frequent length of merozoites is between 6 to 7 microns

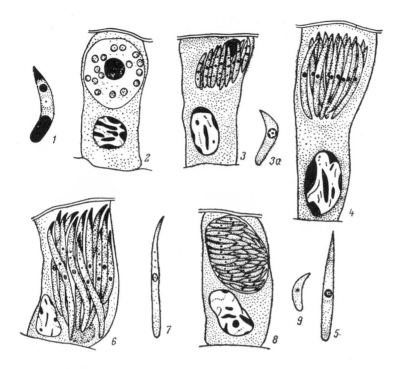

FIG. 13. Asexual generations of *E. nieschulzi* (After Roudabush, 1937). *1*, Sporozoite; *2*, polynuclear first-generation schizont; *3*, mature first-generation schizont; *3a*, first-generation merozoite; *4*, mature second-generation schizont; *5*, second-generation merozoites; *6*, mature third-generation schizont; *7*, third-generation merozoite; *8*, mature fourth-generation schizont; *9*, fourth-generation merozoite.

and 10 to 14 microns. The width of merozoites varies from 0.5 to 3 to 5 microns. There are long and narrow merozoites such as those of *E. intestinalis* and *E. piriformis;* these are 10 to 20 microns long and 0.5 to 1 micron wide. In some species the merozoites may be long and broad, e.g., *E. magna, E. media* and *E. irresidua.* These are 8–12 microns by 1.5–2.5 microns. Short and broad merozoites are 4 to 10 microns long and 2 to 3 microns wide. Such merozoites are those of *E. meleagridis, E. irresidua,* and *E. tenella.* Finally, there may be short and narrow merozoites; *E. nieschulzi,* in addition to long and narrow merozoites, has merozoites 5 to 8 microns long and 1.3 microns wide on the 4th day after infection. In *E. tenella* the merozoites are 3 microns long and 0.5 microns wide (Figs. 11 to 15). A single schizont always forms merozoites of a similar size.

The position of merozoites in schizonts varies. Sometimes the merozoites are situated like the sections of an orange and are united at their
ends with a small residual body lying at the pole. Such an arrangement
is found in schizonts of *E. media, E. irresidua, E. separata* (second generation), and *E. debliecki*. In some species of coccidia the merozoites are
arranged in the schizont like a bunch of bananas. This type of arrangement can usually be observed in second-generation schizonts of *E. intestinalis* and *E. piriformis* from rabbits (Fig. 14). The same is noted in
first-generation schizonts of *E. miyairii* from rats. Finally, in some schizonts of any size the merozoites are randomly arranged. This is ob-

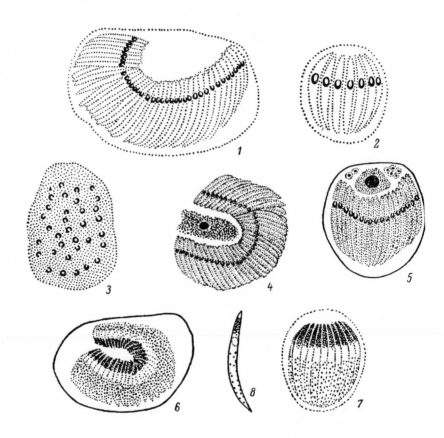

F<small>IG</small>. 14. Second- and third-generation schizonts and merozoites of *E. intestinalis* (After
Kheysin, 1958a). *1* and *2*, Feulgen stain; *3* to *5*, Unna's methyl green-pyronin
stain; *6* and *7*, distribution of glycogen in the segmented schizont; *8*, glycogen in
second-generation merozoite.

served in some third-generation schizonts of *E. magna* and the first generation of *E. adenoeides* and *E. meleagrimitis*. It is possible that such a random arrangement is the result of the movement of the merozoites within the schizont (Pérard, 1924; Kheysin, 1947b). This is all the more probable since some schizonts of the same generation of *E. magna* have merozoites which lie together like a bunch of bananas. When merozoites are situated in an orderly fashion in the schizont they are usually pointed in one direction toward a definite point. This is especially well demonstrated in the large bunches of merozoites within *E. intestinalis* schizonts. Apparently, such a position is caused by the process of merozoite development during schizogony.

In each species of coccidia the merozoites of various generations differ from one another in form and size (Figs. 13 and 14). Sometimes it is possible to observe certain structural differences between them. For example, the first-generation merozoites of *E. tenella* are 2 to 4 microns long, but those of the second generation are 16 microns long, and the third generation are 7 microns long. In *E. magna* the first-generation merozoites are 7–11 microns by 1.5–2.6 microns; second-generation merozoites are 5.3–12.1 by 1.1–2.7 microns; those of the third genera-

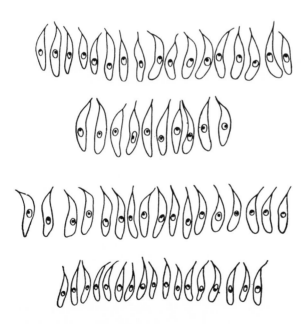

FIG. 15. Merozoites of various generations of *E. magna* (After Kheysin, 1947b).

tion are 5.1–9.5 by 1.0–2.5 microns; and the fourth and fifth generations measure 4.3–7.5 by 1.0–2.3 microns (Fig. 15).

First-generation merozoites of *E. nieschulzi* have the shape of a narrow and very short cigar-like body measuring an average of 8.6 by 1.6 microns. Second-generation merozoites have the same shape, but are longer and thinner (14 by 1.2 microns). Third-generation merozoites are vermiform and reach an average length of 19 microns and a width of 1.2 microns. Fourth-generation merozoites are short and relatively broad (5.5 by 1.3 microns) (Roudabush, 1937). First-generation merozoites of *E. bovis* have a spindle shape and are 8.8 to 15.4 microns long. When observed in vivo, the anterior end has a solid pin-shaped thickening. Second-generation merozoites are almost half the size of first-generation merozoites. The length of live merozoites is 5.7 to 6.8 microns. In fixed and stained preparations, their dimensions decrease to 3–4 by 1.2 microns. The thickening on the anterior end is not visible in such preparations.

Sexual differentiation of merozoites apparently exists in a single generation. As was noted above, Nabih (1938) distinguished three types of schizonts. Likewise, there are three types of merozoites. Protomerozoites are formed in protoschizonts. Eumerozoites are present in euschizonts (the true schizonts), and telomerozoites are formed in teloschizonts. The first two types have only an asexual potential, i.e., they give rise only to new schizonts. The telomerozoites always develop into gamonts after penetration into the host cells. Telomerozoites differ from all others not only structurally, but they also have physiological differences. Like the schizonts, merozoites are seldom referred to by the terms suggested by Nabih. They are most often referred to (with considerable precedence) as merozoites of the first, second, third, or subsequent generations.

In many cases merozoites of various species of coccidia differ from one another just as markedly as the merozoites of different generations within a single species. If two or more species of coccidia are found in one host, it is usually not too difficult to determine the species by their merozoites. For example, in the intestine of a rabbit the broad merozoites of *E. irresidua* are the largest (10–12.5 by 2–4 microns). The merozoites of *E. intestinalis* are long and narrow; this differentiates them from the merozoites of the other species. They are most similar to the merozoites of *E. piriformis* and *E. perforans* but differ from them by their greater length. However, merozoites of different species are not always easily distinguished from one another. Although they differ somewhat in size, the merozoites of *E. media* are very similar to those of *E. magna* (Kheysin, 1947b).

Merozoite structure has been studied very thoroughly with the light microscope. A nucleus is located in the center of the body or nearer to one of the ends. In the long and narrow merozoites of *E. intestinalis,* the nucleus is nearer to the anterior end of the merozoite, but in the merozoites of *E. magna* it is closer to the posterior end. The same is also observed in first-generation merozoites of *E. bovis* (Hammond et al., 1965).

The peripheral layer of the nucleus always gives a distinct Feulgen-positive reaction, but the central part of the nucleus is very weakly stained. Consequently, the greater part of the DNA is concentrated in the peripheral layer of the nucleus (Fig. 14). The karyosome, which stains red with pyronin, is situated in the center of the nucleus or somewhat eccentric to it. After the action of ribonuclease, pyroninophilia disappears. This reveals the presence of RNA in the karyosome. In the nuclei of *E. intestinalis,* the karyosome is absent and RNA is diffusely distributed throughout the nucleus.

Sometimes there is a horseshoe-shaped deposit of chromatin which is highly characteristic of the merozoite nuclei of many coccidian species. This type of chromatin distribution is especially apparent in electron microphotographs in which the chromatin appears as an electron-dense layer at the periphery of the nucleus under its double-layered membrane. Under these conditions, the RNA granules are visible in the center of the nucleus (Kheysin, 1965).

Usually, most species of coccidia have only one nucleus in a merozoite; however, polynuclear merozoites, which have a sickle shape, have been observed in *E. magna, E. irresidua,* and *E. media.* Such merozoites have from two to eight nuclei situated along the body of the merozoites (Waworuntu, 1924; Pérard, 1924; Kheysin, 1940, 1947b), and they are larger than the mononuclear merozoites of the same generation. The binuclear second-generation merozoites of *E. magna* measure 8 by 2 microns, the trinuclear ones were 9 by 3 microns, and the hexanuclear ones 11 by 3 microns. The third generation of binuclear merozoites measures 10 by 3 microns, the tetranuclear ones are 13 by 4 microns, and the octanuclear ones 15 by 5 microns.

The polynuclear merozoites form in small numbers in one schizont. When this happens the merozoites formed always have the same number of nuclei. Their number in *E. magna* and other species from the intestine of rabbits is small when compared with the mononuclear merozoites. In

*E. magna* octanuclear merozoites are present, but in *E. irresidua* no more than four nuclei have ever been observed in merozoites. In *E. intestinalis,* bi- and tri-nuclear merozoites are very rare. Various hypotheses as to their ultimate fate have been advanced. Pérard (1924) thought that polynuclearity was the result of premature division of the nucleus of the merozoite which should ultimately develop into a schizont. Waworuntu (1924) reported that the polynuclear merozoites which he observed in rabbit coccidia produced microgamonts. It is probable that their potential may be varied, and such polynuclear merozoites, after penetration into the host cell, become either schizonts or microgametocytes.

Investigation of merozoites by light microscopy has permitted discovery of only a few structural details. Study of the merozoites of *E. magna* and *E. irresidua* from the rabbit has shown that in smears stained with Heidenhain's iron hematoxylin, highly basophilic spicules are present on the anterior end of the merozoite. They protrude forward about 1.5 to 2 microns (depending on the size of the merozoite) (Kheysin, 1940, 1947b). In merozoites of *E. stiedae,* Waworuntu (1924) observed a rostrum or spicule protruding for a distance of 2 to 6 microns. From this spicule a "fibrilla" extended into the cytoplasm of the merozoite and ended 2 to 3 microns from the nucleus. It is hard to say what this structure is. When the electron microscope was used to study merozoites of *E. magna,* neither the spicule, rostrum, nor fibrilla could be observed on the anterior end of the merozoites. All these structures were observed only when the merozoites were examined in smears; they did not appear in sections. In other species, e.g., *E. intestinalis* or chicken coccidia, such structures again failed to appear. According to Tyzzer et al. (1932) in *E. tenella* and *E. necatrix,* and Roudabush (1937) in *E. nieschulzi* and *E. separata,* only a strongly stained cytoplasmic zone was observed.

When the merozoites of *E. bovis* were treated with protargol, a strongly stained solid cap was present on the anterior end. It formed a sort of ring in the center of which is a pore. From this ring two highly stained rod-shaped structures run down the middle of the merozoite body in the direction of the nucleus (Hammond et al., 1965). It is possible that these structures, visible with light microscopy, correspond to the conoid and paired organelles which appear distinctly in electron microscope investigations. An analogous structure is also present in merozoites of *E. debliecki* (Vetterling, 1966).

Round mitochondria may be observed in the cytoplasm of merozoites of *E. magna* and *E. irresidua* prepared with Kul's stain (Kheysin, 1940, 1947b). These rod-shaped organelles are also observed in second-generation merozoites of *E. tenella* (Gill and Ray, 1957; Nath, Dutta, and Sagar, 1960, 1965). They appear distinct in live merozoites stained with Janus green and when observed with phase-contrast microscopy.

Gill and Ray (1957) observed granules, which stained neutral red, anterior to the nucleus of merozoites of *E. tenella*. They hypothesized that this was part of the Golgi apparatus. Small lipid granules were seen in the same place (Nath et al., 1965).

In early investigations of coccidia there were references to the presence of flagella (Reich, 1913). However, in subsequent investigations these structures have not been observed, and it may be said that merozoites do not have flagella. Also, merozoites do not have centrioles such as were shown on the anterior end of the merozoite drawn schematically in Grassé's *Traité de Zoologie* (1953). Electron microscope investigations of merozoites of several species of coccidia have not revealed the existence of this organelle (Kheysin, 1965; Scholtyseck, 1965b; Sheffield and Hammond, 1965), nor have they revealed any elements of a kinetic apparatus such as a basal body which is analogous to the centriole (Fawcett and Porter, 1954).

Electron microscope research has allowed the discovery of several new details of merozoite structure. At the present the electron microscope has been used to study the merozoites of *E. intestinalis* and *E. perforans* from the rabbit (Mosevich and Kheysin, 1961; Kheysin and Snigirevskaya, 1965; Scholtyseck, 1965b; Scholtyseck and Piekarski, 1965), as well as the merozoites of *E. bovis* from cattle (Sheffield and Hammond, 1965).

The merozoites of *E. intestinalis* are covered with a pellicle consisting of two layers, each having a thickness of 70 Å. Between them is a light space 300 Å thick. The subpellicular fibrils which run in a longitudinal direction are directly under the pellicle. They begin at the anterior end of the merozoite and extend to the posterior end. The number varies from 24 to 30 in various merozoites. In *E. bovis* there are 20 to 22 fibrils. It is difficult to determine the function of these fibrils. They may be some type of supportive material, but the possibility exists that they are the contractile elements of the merozoite. Analogous fibrils are found in other protozoa both in organelles capable of contraction and in those which are not. When merozoites move they not only stretch their bodies but also contract them. It is highly probable that the peripheral fibrils play a definite role in locomotion.

There is a special organelle termed the apical cone or conoid on the anterior end of *E. intestinalis, E. stiedae,* and *E. magna* merozoites. It consists of a broad and solid ring that is shaped like a cone and extends into the cytoplasm behind a thin and short rostrum which is visible only with the electron microscope. The height of the cone is 2700 Å. The cone is attached to the anterior end of the body of the merozoite and may be regarded as an adaptation to aid the merozoite in penetrating the host cell. This cone is similar in appearance to the conoid described in trophozoites of *Toxoplasma gondii* by Ludvik (1958) and the sporozoites of *Plasmodium* (Garnham, Baker, and Bird, 1962, 1963).

In electron micrographs, paired organelles are visible in the anterior third of the merozoites of *E. intestinalis, E. magna, E. stiedae,* and *E. bovis.* They consist of two tubes, each measuring 2 microns long and run parallel to each other from the anterior end of the body in the direction of the nucleus. The anterior ends of these tubes open to the outside in the center of the rings of the conoid. Their diameter at this point is 300 Å. The posterior end is 2000 Å wide and has no aperture. The contents of the tubes are solid material. Analogous organelles have been observed in the sporozoites of *Plasmodium cathemerium, P. falciparum, Lankesterella garnhami,* and in the trophozoites of *Toxoplasma* and *Sarcocystis* (Garnham et al., 1962, 1963). Garnham suggested that this organelle is similar to a gland which secretes a substance that aids the sporozoites in penetration of the host cell. Although this supposition is highly probable it still needs experimental proof. It might be assumed that the delicate fibrils (rhizostyles) running in the direction of the nucleus from the anterior end of the body in the merozoites of *E. magna* and others, which are visible with the light microscope, represent the paired organelles.

At the anterior half of the body of the merozoite, along with the paired organelles, there are numerous solid, coiled structures running from the conoid. Their number reaches 10 or 12 in the different merozoites. The length is approximately equal to that of the paired organelles. Apparently, these structures are analogous to the toxonemes of *Toxoplasma, Sarcocystis, Lankesterella,* and *Plasmodium* (Ludvik, 1958, 1963; Garnham et al., 1962, 1963). It may be assumed that they have a supporting function and strengthen the anterior end of the body of the merozoite.

Thus, the end of the merozoite body has a system of adaptations designed to aid in penetration into the host cell. The system also aids in locomotion. According to my observations on *E. magna* and *E. intestinalis,* movement is brought about by a twisting of the body around the longitudinal axis and contraction, or else by snake-like movements and creeping without any noticeable change in the configuration of the body. Often the merozoite remains in one place but makes several stretching motions after which it quickly advances. This is accompanied by a twisting of the body around the longitudinal axis. Similar movements of merozoites have been observed in *E. bovis* (Hammond et al., 1965). It is still not clear which of the merozoite's numerous fibrils are associated with movement.

Bovee (1965) observed locomotion of the merozoites of *Eimeria* species from the gall bladder of the lizard, *Uma notata*. After a short pause the merozoites began to swim at a speed of 5 to 7 microns per second. They moved 15 to 20 microns, after which they stopped. The rest period did not exceed 3 to 5 seconds, after which they again started swimming forward. During movement there was a twisting of the body and a wavelike stretching. When the advancing movement stopped, a vibration of the forward part of the body continued.

Electron microscope investigations have shown that the cytoplasm of the merozoites of *E. magna, E. intestinalis, E. stiedae,* and *E. bovis* contain mitochondria with numerous short internal tubes (Mosevich and Kheysin, 1961; Scholtyseck and Piekarski, 1965; Sheffield and Hammond, 1965). Such ultrastructure is typical of the mitochondria of many parasitic protozoa.

In *E. bovis* individual elements of the Golgi apparatus with characteristic flattened sacs and pockets have been observed anterior to the nucleus (Sheffield and Hammond, 1965). These structures have also been found in merozoites of *E. perforans* (Scholtyseck and Piekarski, 1965). Apparently, the Golgi apparatus is represented only by individual dictyosomes.

With the use of the electron microscope, a micropore has been observed at the level of the nucleus of *E. intestinalis* merozoites (Kheysin and Snigirevskaya, 1965). This is a small opening with a diameter of 600 Å. It is bounded by two walls of the pellicle which are attached at this point to the cytoplasm. The depth of the micropore is 1500 Å. The micropore is found slightly anterior to the nucleus of third-generation merozoites of *E. magna*. In *E. perforans*, a pore is observed at the posterior end of the merozoite (Scholtyseck and Piekarski, 1965). It is interesting that an analogous ultrastructure has been found in the sporozoites of *Plasmodium* and *Lankesterella* and in the trophozoites of *Toxoplasma, Sarcocystis,* and *Coelotropha durchoni* (Garnham et al., 1963; Ludvik, 1963; Vivier and Henneré, 1965). Garnham et al. referred to this micropore as the micropyle and suggested that the body of the sporozoite of malarial parasites emerges through it after the sporozoite enters the host cell. It is hard to imagine a similar phenomenon with merozoites. Kheysin and Snigirevskaya (1965) hypothesized that the micropore is an ultracytostome similar to the analogous structure found in *Trypanosoma mega* (Steinert and Novikoff, 1960). Further research may reveal the nature of this strange structure. On the basis of electron microscope research, the scheme of coccidial merozoite structure is illustrated in Fig. 16.

Cytochemical research on coccidia of rabbits and chickens has revealed several properties of their metabolism and determined the location of several important chemical components of the cell. Merozoites of *E. magna*, *E. intestinalis*, *E. tenella*, *E. acervulina*, and *E. brunetti* have

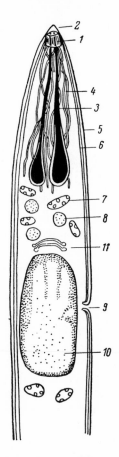

Fig. 16. Diagram of merozoite structure. *1*, Conoid; *2*, plasmatic nose; *3*, paired organelle; *4*, coiled bodies or toxonemes; *5*, pellicle; *6*, peripheral fibrillae; *7*, mitochondria; *8*, protein bodies; *9*, micropore; *10*, nucleus; *11*, golgi apparatus.

a small amount of RNA which is present not only in the karyosome but also in the cytoplasm (Kheysin, 1958a, 1960; Ray and Gill, 1955). During extracellular existence, the merozoites do not synthesize protein and do not grow. However, the presence of RNA in the merozoites is revealed by the fact that after penetration of the host cell, the merozoites (regardless of whether they develop into schizonts or gamonts) undergo rapid growth accompanied by considerable protein synthesis. The presence of a certain amount of RNA in the merozoites makes it possible for growth to begin immediately after penetration into the host cell.

The precursors of RNA, the free purine nucleotides, have not been definitely found in the cytoplasm of merozoites. As a result there is no increase in the RNA content at this stage of the life cycle. However, free purine nucleotides are present in the karyosome of the merozoite nuclei of *E. magna* and *E. coecicola.* It is probable that RNA synthesis may take place here even when the merozoites are outside the host cell.

Grains of volutin have been described in the merozoites of *E. magna.* For a long time the chemical structure of volutin was not known. Subsequent research revealed that volutin is nothing more than RNA (Berghe, 1946). It is quite probable that the basophilic inclusions described in the merozoites of *E. media* by Matsubayashi (1934) were also clusters of RNA.

The cytoplasm of merozoites also contains a certain amount of stored food substances in the form of fat and glycogen. When Sudan III and Sudan black were used as stains, drops of fat were found in the merozoites of *E. magna* and *E. intestinalis* (Kheysin, 1935b, 1947b, 1948). Glycogen has been observed in the merozoites of various species of coccidia. It is chiefly concentrated in the anterior half of the merozoites of all generations of *E. intestinalis* and *E. magna* (Kheysin, 1947b, 1958a, and b). Small clusters of glycogen are also found posterior to the nucleus (Fig. 15). In *E. tenella,* glycogen is found in very small quantities in second-generation merozoites (Gill and Ray, 1954b); according to Edgar et al. (1944), it is not found in merozoites of this species. Glycogen is present in merozoites of *E. acervulina* as individual granules, but none is present in *E. brunetti* (Pattillo and Becker, 1955). There is a considerable amount of glycogen in first-generation merozoites of *E. bovis.* It is concentrated in the center of the body and is nearer to the posterior end (Hammond et al., 1965).

The presence of a large or small supply of glycogen as an internal energy reserve in merozoites is associated with the fact that the merozoites move while in the intestine. As soon as the merozoites penetrate into epithelial cells the supply of glycogen disappears, and energy for the further growth of the trophozoite is obtained at the expense of the influx of nutritive material from the host cell. A new stage of glycogen accumulation then begins in the schizonts or gamonts.

It has been observed in *E. tenella, E. brunetti,* and *E. acervulina* that if the merozoites do not make large migrations and penetrate into the cells near the place of their formation, they possess minimal reserves of glycogen. The merozoites of rabbit and cattle coccidia have large reserves of glycogen which makes it possible for them to undertake considerable migrations in the intestine. In an average-intensity infection, the first generation of *E. magna* localizes about 15 cm from the cecum, and the second generation is found about 30 cm along the length of the colon. On the 8th day after infection, the merozoites of *E. intestinalis* migrate from the small intestine into the cecum and appendix and spread for a long distance along the intestine. The same has been observed with first-generation merozoites of *E. bovis.* They migrate from the small intestine into the large intestine where development of the second-generation schizonts takes place.

In third-generation merozoites of *E. intestinalis,* the amount of glycogen depends to a certain degree on the intensity of infection. The heavier the infection, the more intensively the cells of the epithelium are invaded by schizonts, and the smaller is the amount of glycogen stored in the merozoites. Apparently, this is explained by the carbohydrate deficit which the schizonts experience during development in the intestine. The supply of glycogen in the merozoites of the later generations of *E. magna* and *E. intestinalis* proved to be higher than that in first-generation merozoites. This may be associated with the fact that the first-generation merozoites do not resettle a long distance away, but third- and fourth-generation merozoites migrate along the small intestine not only in the direction of the stomach but also in the direction of the large intestine as far as the cecum.

The merozoites have aerobic respiration which is the most efficient source of energy for this stage of development. The carbohydrate substrate necessary for this type of respiration is accumulated during schizogony. Although succinodehydrogenase activity has not been found in schizonts which have anaerobic carbohydrate respiration, it can be demonstrated in the merozoites. In the merozoites of *E. magna* and *E. intestinalis,* the granules of mono- and diformazan are located along both sides of the nucleus, and often the greatest concentration is observed anterior to the nucleus where there is an especially large number of mitochondria (Beyer, 1962).

The cytoplasm of the merozoites of *E. magna* and *E. intestinalis* stains poorly when Chevremont and Frederik's blue stain is used; this indicates the presence of thiol compounds. The karyosome and cytoplasm also stain. By the use of Burnett and Seligman's method, SH-groups, which are associated with protein, have been demonstrated in the cytoplasm. The appearance of thiol compounds (which are not present in the schizonts) in the merozoites is apparently associated with the transition of the merozoites to anaerobic respiration.

Alkaline phosphatase has been discovered only in the nuclei of *E. magna* merozoites (Beyer, 1960). It is concentrated in the karyosome, but the nucleoplasm stains very weakly. This was also found in the merozoites of *E. tenella* (Gill and Ray, 1954a) and *E. stiedae* (Dasgupta, 1959). In *E. intestinalis* the reaction to alkaline phosphatase is significantly weaker than in *E. magna* (Beyer, 1960). Acid phosphatase activity in *E. magna* and *E. intestinalis* has been observed in both the nucleus and cytoplasm, especially in the anterior half of the merozoite. Protein granules are also present there and stain blue when treated with mercury-bromphenol blue. These granules do not give a positive reaction to acid mucopolysaccharides.

An increase in the amount of glutathione and free cysteine, as well as a rise in the concentration of SH- and SS-groups of protein along with activation of the dehydrogenase of succinic acid, demonstrates the high intensity of the oxidation-reduction processes in the merozoites of *E. magna* and *E. intestinalis* (Beyer, 1962).

### THE SEXUAL PROCESS

The development of male and female gametes always begins with a mononuclear trophozoite which develops from the merozoite after the latter has penetrated the host cell. The trophozoites, from which the gamonts or schizonts are formed, are very difficult to distinguish from one another by present methods of study. The sexual process in coccidia is broken down into two periods: (1) the gametogenesis period during which formation of male and female gametes occurs and (2) the fertilization period and subsequent formation of the zygote.

Gametogenesis in coccidia is characterized by the fact that one comparatively large macrogamete always develops from one macrogamont, but a multitude of small, motile, flagellated microgametes are formed from one microgamont or microgametocyte. During macrogametogenesis there is an accumulation of nutritive and "plastic" substances.

Microgametogenesis is accompanied by the growth of the microgameto-cyte, division of the nuclei, and subsequent differentiation of the gametes.

## Microgametes and Microgametogenesis

The cytoplasm of a mononuclear trophozoite which begins to develop into a microgametocyte contains little glycogen and RNA, and the nucleus has the typical merozoite structures with a peripheral distribution of chromatin. However, unlike the nuclei of merozoites, the chromatin in the mononuclear microgamete is distributed evenly along the surface of the nucleus and does not form a sickle-shaped cluster such as is observed in merozoites of many species. According to Dobell (1925), microgametocytes of *Aggregata* have a small centriole next to the nucleus which takes part in mitotic division. A centriole has not been observed in the microgametocytes of other coccidia.

The process of microgametogenesis can be divided into two periods. The first is the period of gametocyte growth and is accompanied by a reproduction of nuclei. Growth of the gametocyte and division of the nucleus leads to the formation of a polynuclear microgametocyte with a characteristic peripheral distribution of nuclei. The second period is one of differentiation and formation of microgametes (Kheysin, 1940, 1947b, 1958a; Scholtyseck, 1965a).

During the period of gametocyte growth, the volume increases many times. The young microgametocytes of various species have a diameter of approximately 3 to 5 microns. By the end of the first period, the microgametocyte reaches its maximum size. For example, in *E. perforans* the microgametocyte measures 12 to 15 microns; in *E. media,* 15–25 microns by 9–17 microns; *E. coecicola,* 20–22 microns by 12–17 microns; *E. magna,* 17 to 40 microns; *E. intestinalis,* 22 to 27 microns; and in *E. irresidua,* 30–60 microns by 30–50 microns. In *E. tenella* the small microgametocytes are 12.4 by 8.7 microns, while in *E. mitis* they are even smaller (9.5 to 13.5 microns). The dimensions are about the same for microgametocytes of turkey coccidia. Very large microgameto-cytes were found in *E. auburnensis* from cattle. They reached a mean size of 85 by 65 microns (Hammond, Clark, and Miner, 1961). The microgametocytes of *E. leuckarti* from the horse are even larger. According to Kupke (1923) they reach a size of 300 by 170 microns. In *E. cameli,* a parasite of camels, the microgametocytes are 200 by 150 microns (Enigk, 1934). Thus, the volume of microgametocytes increases approximately 80 to 400 times during growth.

Each species of coccidia is characterized not only by absolute but also by relative sizes of microgametocytes. In some cases their dimensions do not exceed those of mature macrogametes or are even smaller. Examples are the microgametocytes of *E. intestinalis, E. media, E. nieschulzi, E. perforans, E. separata, E. mitis, E. tenella, E. meleagrimitis, E. adenoeides,* and others. In other cases the microgametocytes are larger than the macrogametes. This can be observed, for example, in *E. irresidua* and *E. magna* from the rabbit, *E. brunetti* and *E. maxima* from chickens, and *E. auburnensis* from cattle.

The larger the microgametocyte, the greater the number of nuclei within it. In the large microgametocytes of *E. irresidua* or *E. auburnensis,* more than a thousand nuclei are present, but in the smaller gametocytes of *E. magna* there are never more than a thousand, and in *E. intestinalis* there are about 500. In the microgametocytes of *E. maxima,* which are 24 to 48 microns (mean 31 microns), as many as 600 gametes are formed.

On the basis of their structure, two forms of gametocytes are recognized. During growth of the first type, the nuclei are situated in one row at the periphery around the central mass of cytoplasm which ultimately develops into a residual body when the microgametes are formed. These are the so-called monocentric microgametocytes. *Eimeria perforans, E. intestinalis, E. media, E. adenoeides, E. meleagrimitis, E. separata, E. tenella,* and *E. mitis* are examples of this type of development. In the second type of microgametocyte, as the central mass grows it separates into several parts and the nuclei become situated around each part. An impression of several centers of microgamete formation results. There may be 2 or 3 or even 5 to 6 such parts. These centers are sometimes called cytomeres or microgametoblasts (Tyzzer, 1929). Therefore, such a microgametocyte is termed polycentric. After formation of the microgametes, each gametoblast becomes a residual body. Among the rabbit coccidia, *E. irresidua* and *E. magna* have polycentric microgametocytes. They are also formed in the coccidia of chickens, e.g., *E. maxima.* Microgametoblasts have also been described in the microgametocytes of *Caryospora* and *Isospora felis* (Reichenow, 1953; Lickfeld, 1959).

The nuclei of the gametocytes have a structure similar to that of the schizonts. The peripheral zone of chromatin gives a distinct Feulgen-positive reaction. A karyosome is often visible in the center of the nucleus. As the nucleus divides, the karyosome becomes smaller and disappears by the end of the growth period (Kheysin, 1947b, 1965). The young gametocytes have larger nuclei than the mature ones have. As the number of nuclei increases, the size decreases to approximately one-half that of nuclei in mononuclear gametocytes. Nuclear division is by mitosis. In some coccidia (e.g., *Aggregata*) the chromosomes are clearly visible. During the metaphase stage of *Aggregata*, it is easy to distinguish the six chromosomes lying along the equator of the spindle (Dobell, 1925).

After the nuclei are located at the periphery of the gametocyte, growth of the latter ends, and division of the nuclei also ends. With the use of light microscopy, it is possible to observe that differentiation of the microgametes begins with reconstruction of the nuclei. In *E. magna* and *E. intestinalis*, the diameter of the nuclei decreases and the chromatin occupies a greater proportion of the nucleus. Then the flattened nuclei, which are filled with chromatin, lengthen somewhat and assume the shape of a short flat comma. Such nuclei later stretch even further and retain the comma shape. The final formation of microgametes takes place after the nucleus stretches into the shape of a narrow and delicate comma. In such microgametes observed with light microscopy, the nuclei become located on the surface of the residual body and then move away from it; due to the presence of two flagella, the microgametes move into the space between the residual body and the edge of the gametocyte. The nuclei of the microgametes always show up well in a Feulgen-positive reaction.

Such formation of microgametes is more or less characteristic of all representatives of the family Eimeriidae. In *Aggregata*, the process of microgamete formation begins when each nucleus becomes situated at the periphery of the gamont; a small amount of cytoplasm separates itself, and along with a nucleus, rises up from the gametocyte in the form of a small bud (Fig. 17). The nucleus then becomes compressed and flattened. A centriole, from which a pair of flagella 15 microns long develops, lies terminally near the nucleus at the distal end of the now forming microgamete. The rudiment of the microgamete then stretches and for some time preserves the connection with the residual body of the microgametocyte with its posterior end. When the microgamete reaches 34 to 45 microns in length [sic], it separates from the residual body (Fig. 17).

With light microscopy it is possible to observe two posteriorly directed flagella in all microgametes. Such microgametes are very similar to the flagellated *Bodo*. In drawings of *Aggregata* microgametes by Dobell, there are two anterior flagella and also a third flagellum which is directed along the body toward the posterior to form something similar to an undulating membrane (Fig. 17).

The microgametes, which are shaped like a stretched membrane in

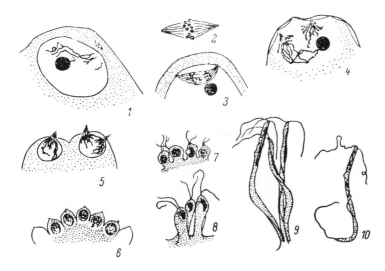

Fig. 17. Microgametogenesis in *Aggregata eberthi* (After Dobell, 1925). *1*, Preparation for nuclear division (centriole visible); *2* to *3*, metaphase and anaphase; *4*, end of division; *5*, microgametocyte nucleus at the end of the growth period; *6* to *9*, formation of the microgametes; *10*, microgamete.

different coccidia, are approximately 5 to 8 microns long and 0.5 to 2.2 microns wide. In some species, very small microgametes are found. For example, in *E. perforans* the length of the gametes is 3 microns and the length of the flagella is 10 microns. In *E. separata* there are even smaller gametes 1.8 to 2.7 microns long and 0.5 microns wide. In *E. magna* the length is 3 to 5 microns and the width is 0.6 to 0.8 microns; the length of the flagella is 15 microns. In *Cyclospora caryolytica* the length of the gametes is 8 microns and the length of the flagella is 16 microns. Thus, the flagella are 2 or 3 times as long as the microgamete body. The movements of the flagella provide forward locomotion of the microgamete in a narrow pathway.

Electron microscope investigations of the microgametes of *E. magna*, *E. intestinalis*, and *E. perforans* (Kheysin, 1963, 1965; Scholtyseck, 1965a) have clarified a good deal about their structure and the process of formation during microgametogenesis. The nucleus occupies the greater part of the microgamete. The anterior part, which consists of approximately one-third of the entire length of the gamete, is the cytoplasmic portion of the gamete. A large mitochondrion, with short tubules lying in rows under its external membrane, is located in front of the

nucleus. Two basal bodies are present in the anterior part of the microgamete. They are directed forward, but the flagella emerge from the
body pointed toward the posterior end (Fig. 18). The apical (terminal)
end of the microgamete is flattened to form a perforatorium. The microgametes move with this end forward. Two flagella have been observed in
microgametes of *E. magna* and *E. intestinalis,* but Scholtyseck observed
three flagella in *E. perforans.* Two flagella run back freely, and the short
third one is also pointed backwards but is connected to the body of the
microgamete for a short distance. It may be that a third flagellum will
also be found in other species of coccidia. However, detailed and accur-

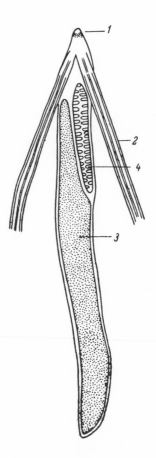

Fig. 18. Diagram of the microgamete structure. *1,* Apical cap; *2,* flagellum; *3,* nucleus; *4,*
mitochondrion.

ate research is needed to bring to a final conclusion the question of the number of flagella on the microgametes.

The process of microgamete formation has been studied in the same three species (Kheysin, 1965; Scholtyseck, 1965a). Electron microscopy has shown that an important part of microgametogenesis is the formation of the basal bodies and flagella during the first period of microgametogenesis (Kheysin, 1965). At an early stage of microgametocyte development, the nuclei are arranged in a random manner in the cytoplasm; at this time it is impossible to observe any elements of the basal body or centriole. The rudiments of the basal bodies appear only when the nuclei in the gametocyte occupy a peripheral position. At this time the rudiments of the basal bodies are formed between the nuclei and the pellicle as nine short tubular fibrils with a diameter of 200 to 250 Å. It is highly probable that they arise *de novo* or from the fibrillar tubular structures of the cytoplasm which become organized into the structures of the basal body. Later, as the nuclei divide, the pellicle forms small buds on the surface of the gametocyte near the rudiments of the basal bodies; these are the rudiments of the flagella. The buds are penetrated by the tubular fibrils of the basal bodies, after which the rudiment begins to lengthen and assume the typical form of the flagella. The axial cylinder consists of nine pairs of peripheral fibrils and one pair of central fibrils. At the end of the first period of gametogenesis, short flagella are visible on the surface of the gametocyte. Near each nucleus, in addition to the basal bodies, are large mitochondria which are apparently formed by the merging of small mitochondria; these are similar to the accessory nuclei (Nebenkern) of the spermatids of various insects (Vil'son, 1936).

The second period of microgametogenesis begins with the appearance of the rudiments of the microgametes on the surface of the microgametocyte. These are protuberances containing nuclei, mitochondria, and basal bodies. Although light microscopy reveals that the second period begins with a change in nuclear structure, the electron microscope shows the rudiments of the microgametes as they form. Further microgamete development is associated with a lengthening of the newly formed rudiment, nucleus, and flagella. The anterior end of the microgamete is attached to the cytoplasm of the gametocyte. The cytoplasm later becomes a residual body (Kheysin, 1965). According to Scholtyseck (1965a), the nucleus of the rudimentary microgamete of *E. perforans* is at first twisted into the form of a comma, but it then straightens out and stretches as the gamete develops and increases in size.

Separation of the microgamete from the residual body takes place after the flagella reach their maximum size. The entire microgamete buds off of the surface of the gametocyte. During the first period of microgametogenesis, a considerable amount of RNA accumulates in the cytoplasm of the growing microgametocyte. This can be stained with pyronin or gallocyanine. A negative reaction occurs after the preparation is treated with ribonuclease. The presence of RNA is associated with gametogenesis during a period of uninterrupted nuclear division and active synthetic processes which nourish the growth of the gametocyte. RNA has also been observed in the microgametocytes of *E. stiedae* (Dasgupta, 1961b).

There are few nutritive substances in the microgametocytes of *E. magna* and *E. intestinalis*. A small amount of glycogen can be found, but after formation of the microgametes it passes into the residual body. No carbohydrates have been observed in the microgametocytes of *E. acervulina* and *E. brunetti* (Pattillo and Becker, 1955) or in *E. tenella* (Edgar et al., 1944).

During the second period of gametogenesis no changes in RNA content are observed. It is interesting that the residual body retains large supplies of RNA after formation of the gametes. It is difficult to say what the significance of this phenomenon is. A small amount of RNA is observed in the microgametes of *E. tenella* (Ray and Gill, 1955). The microgametes of *E. tenella* also give a positive reaction for carbohydrates (Gill and Ray, 1954b and c). However, Pattillo and Becker (1955) did not observe carbohydrates or RNA in the microgametocytes of *E. acervulina* or *E. brunetti*. The microgametes are very small, so it is not surprising that small amounts of RNA and carbohydrates might be missed by investigators. The cytoplasm of the microgametocytes of *E. intestinalis* and *E. magna* gives a weak reaction to proteins when treated with mercury-bromphenol blue, but the nuclei and the microgametes stain a deep blue.

In the early stages of gametogenesis, free purine nucleotides are present in the cytoplasm. As the reaction to the free purine nucleotides decreases, the gametes grow. A positive reaction is observed both in the cytoplasm and in the karyosome of the nucleus. The microgametes of *E. magna* yield a negative reaction to free purine nucleotides (Beyer, 1962).

At all stages of gametogenesis, the SH-groups associated with proteins can be found in the cytoplasm. However, by the time the first period is completed, the reaction of the cytoplasm to the SH-group noticeably weakens. On the other hand, the amount of disulfide protein groups in the karyosome of the numerous nuclei increases.

In studies of *E. magna, E. intestinalis* from rabbits, and *E. tenella*

from chickens (Beyer, 1960; Gill and Ray, 1954a), it has been established that in the growing microgametocytes, alkaline phosphatase activity was present only in the nuclei and primarily in the karyosomes (e.g.,
*E. magna*). After formation of the microgametes, the nuclei gave a distinct positive reaction to alkaline phosphatase, but the residual body did
not contain the enzyme. The nucleus of the developing gametocyte also
contained acid phosphatase which was noticeable only in the karyosome.
It has been impossible to find succinodehydrogenase activity in growing
microgametocytes (Beyer, 1962).

The presence of a large supply of RNA during the second period of
gametogenesis is associated with the fact that the flagella are growing
rapidly at this time; this is possible only during intense protein synthesis.
In *E. magna*, about 1000 microgametes are formed from each microgametocyte. Each contains two flagella which measure about 10 microns.
Consequently, a total of 2000 flagella with a total length of 2 cm are
formed. For the formation of such a quantity of flagella, it is undoubtedly necessary to have a large amount of protein, the synthesis of which
takes place within the microgametocyte.

The energy for locomotion of the microgamete is apparently associated with the large mitochondrion located anterior to the nucleus near
the basal bodies. It is possible that the energy source is a small supply of
glycogen, which in the opinion of Ray and Gill, is found in the microgametes of *E. tenella*. The gametes move in a limited area, and therefore,
a relatively small energy source is sufficient for microgamete movement.
They penetrate rather rapidly into the macrogametes.

## Macrogametes and Their Development

The mononuclear trophozoites from which the macrogametes grow,
differ at the start of their growth from the microgametocytes or schizonts. At the very earliest stages of development the nucleus of the
future macrogamete rapidly and disproportionately increases in size. It
also loses the ability to react positively to the Feulgen test. In *E. magna*
this phenomenon takes place when the trophozoite reaches a diameter of
6 microns. At the same time, basophilia (which depends on the presence
of RNA) increases in the cytoplasm.

Further growth of the macrogamete is associated with an accumulation of large amounts of nutritive and plastic materials for use during
the metagamic period of development. During growth of the macrogamete, a significant increase in the size of the cytoplasm and nucleus
occurs. The observations of Scholtyseck (1963a) on *E. maxima* show

that during the early stage of development the nucleus grows more rapidly than the cytoplasm. At a later period the cytoplasm somewhat outgrows the nucleus or at least maintains the same growth. The same has been observed in *Aggregata octopiana, A. eberthi,* and *Klossia helicina.* As a result, the nuclei of young macrogametes occupy less than one half the space of the entire macrogamete, but in adult macrogametes the nuclear-cytoplasmic ratio is about even. Consequently, there are two phases of macrogamete growth. During the first phase, growth of the nucleus overtakes that of the cytoplasm, and during the second they grow in proportion.

At an early stage of growth the macrogametes are usually spherical but when maximum growth is reached they may become oval, ellipsoidal, or some other shape which will determine the shape of the oocysts formed from the fertilized macrogamete.

The size of macrogametes at the time of fertilization varies considerably and fluctuates between 10–15 and 35–90 microns. The smallest macrogametes form in those species which develop in the striated border of intestinal epithelial cells, i.e., almost extracellularly. For example, in *Cryptosporidium parvum* from the intestines of mice, the macrogametes have a diameter of no more than 4.5 microns prior to fertilization. In *E. coecicola, E. perforans, E. intestinalis, E. meleagrimitis,* and others, the larger macrogametes develop in epithelial cells. The diameter of completely mature macrogametes may be less than that of the epithelial cell itself. In others the macrogamete occupies the entire epithelial cell and even stretches it, e.g., *E. nieschulzi, E. piriformis,* and others.

The dimensions of the epithelial cells limit the growth of macrogametes to a certain degree, so that the length of the macrogamete usually corresponds to the size of the cell. In cases in which two macrogametes are present in one cell (this sometimes occurs in severe cases of *E. intestinalis* in rabbits) their size is correspondingly smaller than that when a single macrogamete develops.

The largest macrogametes develop in connective tissue (*E. bovis* [sic], *E. leuckarti, E. gilruthi,* and others) or in epithelial cells which are connected to the underlying connective tissue (*E. magna, E. irresidua, E. maxima,* and others). The macrogametes of *E. magna* and *E. irresidua,* which develop in the subepithelium, are longer than the macrogametes of *E. intestinalis, E. piriformis,* and *E. perforans* which develop only in epithelial cells (Kheysin, 1947b, 1948). The very large macrogametes of *E. leuckarti* are formed in the subepithelium, apparently in connective tissue cells in the intestines of horses. Their length reaches 70 to 80 microns. The macrogametes of *E. cameli* are approximately the same size. They develop in epithelial cells which are linked to the connective tissue.

As was mentioned earlier, during the growth of macrogametes an increase in the size of the nucleus takes place. When this occurs the nucleus undergoes no division, and no processes occur which might be interpreted as a reduction or ejection of chromatin. This should be kept in mind, because in some of the older papers by Schaudinn (1900), Reich (1913), and others it is stated that during growth of the macrogamete an ejection of part of the chromatin from the nucleus into the cytoplasm occurs. The nucleus of the macrogamete, as has been established by all the investigators studying the process of gametogenesis by modern cytochemical methods, does not undergo reduction of chromatin and does not eject chromatin into the cytoplasm. The cytoplasm has never been found to contain any Feulgen-positive bodies or granules which might be interpreted as part of the chromatin from the nucleus (Kheysin, 1947b, 1958a, 1960; Pattillo and Becker, 1955). However, some investigators (Ray and Gill, 1955; Dasgupta, 1959) have found a small number of Feulgen-positive granules in the cytoplasm of macrogametes and oocysts of *E. tenella* and *E. stiedae*. Probably these are of viral or bacterial origin.

During growth of the macrogamete, the nucleus remains Feulgennegative. This has been observed in rabbit coccidia (Zasukhin, 1935; Kheysin, 1940, 1947b, 1958a, 1960) and in bird coccidia (Pattillo and Becker, 1955; Ray and Gill, 1955; Tsunoda and Itikawa, 1955, and others). In this respect the nuclei of the macrogametes resemble the nuclei of growing oocytes from various animals. The absence of Feulgen-positive material in the nuclei does not prove the absence of DNA from the nuclei. After the use of Unna's method with methyl green-pyronin, the nuclei of *E. intestinalis* and *E. magna* macrogametes showed a weak reaction at the peripheral zone of the nucleus of growing macrogametocytes (Kheysin, 1958a, 1960). This shows the presence of DNA in the nucleus; however, it is found in such a state (high-polymer DNA) that it does not show up in the Feulgen reaction. The Feulgen method reveals only low-polymer DNA in fixed preparations. A negative reaction to the Feulgen test may also be the result of the fact that the concentration of DNA in the large nucleus of the macrogamete is too low [to give a positive reaction. Ed.]. Thus, DNA is always present in the nuclei at all stages of the developmental cycle of coccidia, and it does not disappear during macrogametogenesis.

The nucleoplasm stains an even, light green color when treated with fast green during all stages of macrogamete growth; this shows an alkaline medium of histone. The nuclei of macrogametes of various coccidia often have a half-moon shaped aggregation which stains deeply with nuclear stains. In the nucleus of macrogametes of *E. magna* and *E. intestinalis*, this clustering gives a positive reaction to histone and is apparently due to concentration caused by fixation.

The nucleus of the macrogamete is characterized by the presence of a karyosome which has a distinct basophilia; this is the result of the presence of RNA (Ray and Gill, 1955; Pattillo and Becker, 1955; Tsunoda and Itikawa, 1955; Kheysin, 1958a, 1960). The karyosome usually is fairly large. In young macrogametes of *E. magna* the diameter of the karyosome is usually 0.5 to 0.8 microns, but in larger macrogametes which measure 5 to 20 microns, the karyosome reaches 1.2 to 1.5 microns. It reaches a maximum diameter of 2 microns in gametes measuring 15 to 18 microns. When growth of the macrogamete ends at the moment of fertilization, the karyosome has decreased to 1 to 1.5 microns.

Using cytochemical methods, Beyer (1962, 1963a) demonstrated that the karyosomes of *E. magna, E. intestinalis,* and *E. coecicola* contained not only RNA but also free purine nucleotides, thiol protein groups, glutathione, free cysteine, and acid and alkaline phosphatase. Because these substances are of direct importance in protein synthesis, it is possible to consider that the karyosome of the macrogamete nucleus participates directly in the process. During growth of the macrogamete and increased synthesis of proteins, carbohydrates, and lipids, the karyosome increases proportionately in size. When synthesis is curtailed and redistribution of plastic substances takes place to form the wall, the karyosome diminishes correspondingly.

Two substances necessary for protein synthesis are present in the cytoplasm of macrogametes. The cytoplasm contains a large amount of RNA. Photometric research by Beyer and Ovchinnikova (1964) on macrogametogenesis of *E. intestinalis* and *E. magna* has shown that the amount of RNA increases as the macrogamete grows and reaches a maximum size before fertilization (Table 1).

In preparations stained by Unna's method, the cytoplasm of the young macrogametes of *E. magna* and *E. intestinalis* is more pyroninophilic than that of large macrogametes. The impression is created that the amount of RNA decreases with the growth of the macrogamete. For example, using visual observations, Scholtyseck (1963a) noted a decrease of RNA in the macrogametes of *E. maxima* during growth. Photometric studies have shown that during growth of macrogametes there is a change in the concentration of RNA per unit of volume. When

TABLE 1
Change in amount of RNA macrogametogenesis of *E. magna* and
*E. intestinalis* in relative quantities (after Beyer and Ovchinnikova, 1965)

| Size of macrogamete in $\mu^3$ (V) | | Amount of RNA (q) | | Concentration of RNA (q/V) | |
|---|---|---|---|---|---|
| *E. magna* | *E. intestinalis* | *E. magna* | *E. intest.* | *E. magna* | *E. intest.* |
| 24.63 | 5.75 | 16.32 | 3.67 | 0.66 | 0.63 |
| 63.87 | 20.30 | 41.38 | 15.12 | 0.65 | 0.74 |
| 103.00 | 39.20 | 43.66 | 30.04 | 0.42 | 0.77 |
| 161.10 | 62.73 | 74.27 | 38.70 | 0.47 | 0.62 |
| 229.50 | 92.50 | 134.00 | 54.00 | 0.48 | 0.58 |

the volume of the macrogamete increases rapidly, the concentration of RNA decreases, but its absolute quantity increases. After the end of macrogamete growth the RNA level does not change, but after fertilization the amount decreases. In newly formed oocysts which are located inside cells, the RNA level remains unchanged. During growth of the macrogamete, the large increase in RNA content of the cytoplasm and karyosome is associated not only with growth, but also with the formation of a large number of protein and glycoprotein granules which are used for construction of the oocyst walls. Cessation of RNA synthesis in the oocysts is apparently associated with the fact that the wall stops exchange with the host cell, and the RNA content in the zygote is stabilized. This amount of RNA provides for further synthetic processes which are associated with formation of the sporozoites and the sporocyst wall.

In the young macrogametes of *E. intestinalis,* there is a concentration of free purine nucleotides, chiefly in the karyosome, but some are located in the cytoplasm. As the macrogametes grow, the reaction to free purine nucleotides gradually decreases in the cytoplasm but remains stable in the karyosome. Prior to fertilization of the macrogametes, free purine nucleotides are located in the protein granules situated along the periphery of the macrogametes (Beyer, 1963).

Other substances are needed for protein synthesis, such as thiol compounds and the phosphomonoesterases in the cytoplasm of the young and growing macrogametes. The highest concentration is found in young macrogametes.

The macrogametes begin to accumulate glycogen very early (Fig. 19). The gametes of *E. magna* and *E. intestinalis,* which measure 9 to 10 microns, contain no glycogen, but the gametes measuring 12 to 15 microns have small glycogen granules around the nuclei. An analogous situation has been reported in macrogametes of *E. acervulina* and *E.*

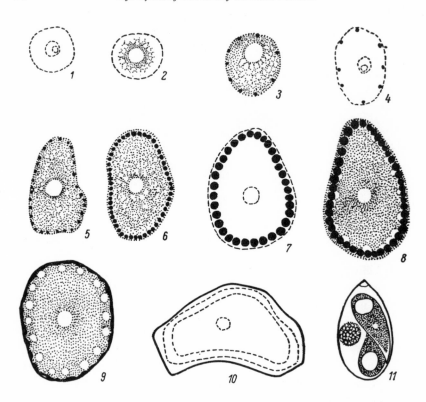

FIG. 19. Development of *E. intestinalis* macrogametes (After Kheysin, 1958a). Periodic
acid-Schiff (PAS) stain for polysaccharides; *1* to *3*, Early stages of the development of macrogametes; *4* to *7*, PAS stain after the action of diastase; *5, 6,* and *8,*
large macrogametes with glycoprotein granules; *9,* formation of the oocyst wall
from the granules; *10,* oocyst stained with PAS after treatment with diastase; *11,*
a sporocyst stained with iodine. The small granules are glycogen and the large
granules are glycoprotein.

*brunetti* (Pattillo and Becker, 1955) and *E. tenella* (Edgar et al., 1944;
Tsunoda and Itikawa, 1955). As the gametes grow the amount of glycogen increases, and in the large gametes of *E. magna* (Kheysin, 1958), *E.
falciformis, E. stiedae* (Giovannola, 1934), and other species, the entire
cytoplasm was filled with this reserve nutritive material which is used by
the macrogamete in glycolytic respiration as a source of energy for the
subsequent processes of sporogony.

In the young macrogametes of *E. magna, E. intestinalis,* and *E.
media,* small drops of lipids appear along with the glycogen during the

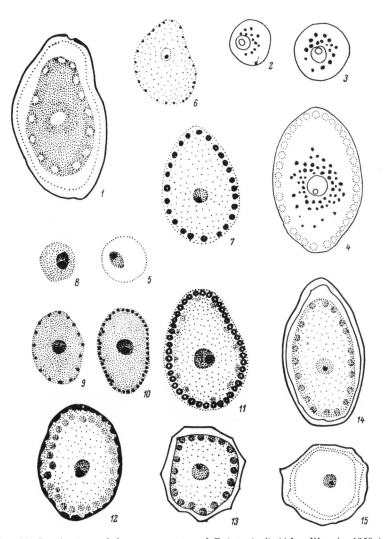

FIG. 20. Development of the macrogametes of *E. intestinalis* (After Kheysin, 1958a). *1*, Oocyst from the intestinal cavity (periodic acid-Schiff reaction; external wall stained red, protein granules not stained); *2* to *4*, accumulation of lipid granules during different stages of macrogamete development; *5* to *7*, macrogametes at various stages of growth and the oocyst *(15)* in transverse section (stain for mucopolysaccharides after Hale; peripheral granules and external wall blue-green; internal wall not stained); *8* to *12*, macrogametes at various stages of growth (Sulema-bromphenol blue stain; peripheral glycoprotein granules are dark blue, protein granules are bright blue); *13* to *14*, formation of the external and internal oocyst walls (Sulema-bromphenol blue stain; protein granules in cytoplasm).

process of macrogamete growth and concentrate around the nucleus (Fig. 20). In addition to glycogen and lipid droplets, the young macrogametes of *E. magna* and *E. intestinalis* also have small granules which give a strong positive reaction for protein when the mercury-bromphenol blue test is used. As the gametes grow, the number and size of these granules increases, and they become arranged at the periphery of the macrogametes. Analogous granules have been observed in *E. acervulina, E. brunetti,* and *E. tenella* (Pattillo and Becker, 1955; Ray and Gill, 1955). Apparently they may be present in all macrogametes.

Along with the protein granules, which reach a size of 12 to 15 microns [sic] in the macrogametes of *E. intestinalis* and *E. magna,* small granules (up to 0.2 microns) also begin to form and give a positive reaction for protein when stained with mercury-bromphenol blue and for polysaccharide with periodic acid-Schiff's reagent (PAS) (Figs. 19 and 20). After treatment with thialine or hyaluronidase, the red color from the PAS reaction does not disappear. The granules are not digested by diastase, nor do they display gamma-metachromasia or react for lipids. Thus, these glycoprotein granules contain neutral mucopolysaccharides. As the gametes grow, they increase in size and finally become distributed along the periphery of the gamete (Figs. 19 and 20). Somewhat deeper are the protein granules which ultimately form the internal wall of the oocyst. The glycoprotein granules form the external wall (Kheysin, 1958a, 1960). The same glycoprotein granules are also present in the macrogametes of *E. acervulina* and *E. brunetti* (Pattillo and Becker, 1955). According to Ray and Gill (1955), the peripheral granules in *E. tenella* give a positive reaction for RNA, but Tsunoda and Itikawa (1955) found no RNA in granules of this species. The same negative results were also obtained for the macrogametes of *E. acervulina* and *E. brunetti* (Pattillo and Becker, 1955) and for *E. magna* and *E. intestinalis* (Kheysin, 1958a, 1960).

Before the use of cytochemical methods, various granules in the macrogametes were referred to as chromatoid and plastinoid (Reich, 1913; Hosoda, 1928, and others). Apparently, the former corresponded to protein and the latter to glycoprotein granules. According to Beyer (1963a), proteins containing sulfhydryl groups participate in the formation of both types of granules in *E. magna* and *E. intestinalis.*

Löser and Gönnert (1965) reported that young macrogametes of *E. stiedae, E. falciformis, E. travassosi,* and *Cyclospora caryolytica* have homogenous bodies around the nucleus which are stained red by azan. They also stain with malachite green, which indicates that they are secretory granules which ultimately take part in the formation of the oocyst wall. These granules correspond to the plastinoid or mucopolysaccharide granules. These authors reported that the plastinoid granules might represent the mucous secretions which form around the nucleus.

Nath, Dutta, and Sagar (1961) used a method of freezing tissue for subsequent tests for lipids; they found that the peripheral granules of the macrogametes of *E. tenella* contained lipids which enter into the structure of the future oocyst wall. However, Gill and Ray (1954), Pattillo and Becker (1955), and Kheysin (1958a, 1960) failed to observe lipids in these granules. It should be pointed out that these authors did not use the freezing method and possibly lost the small quantity of lipids which were present in the peripheral granules.

Ray and Gill found acid mucopolysaccharides of the hyaluronic acid type on the surface granules of *E. tenella.* Tsunoda and Itikawa (1955) also reported that the glycoprotein granules contain hyaluronic acid, as well as chondrioitinsulfuric acid. Actually, these granules in the macrogametes did stain green when Hale's method was used with dialyzed iron for hyaluronic acid. However, after treatment of the preparation with hyaluronidase this color was maintained (Kheysin, 1958a, 1960). Braden (1955) noted that the color in the Hale test was not specific for acid mucopolysaccharides. This method also reveals neutral mucopolysaccharides which are also present in the granules (Fig. 20).

In order to find acid mucopolysaccharides in the macrogametes of *E. intestinalis* and *E. magna,* a method of staining with toluidine blue has been used; the process produces gamma-metachromasia (Kheysin, 1958a, 1960). However, gamma-metachromasia is eliminated by ribonuclease, hyaluronidase, or hot water. It may be assumed that the granules appearing in the preparations do not serve as an indicator of the presence of acid mucopolysaccharides, but represent a false metachromasia or the presence of highly polymerized carbohydrates. Staining with Alcian blue for acid mucopolysaccharides also proved negative. The peripheral granules did stain blue, but the same color was observed in control preparations treated with hyaluronidase.

Throughout the growth of the macrogamete, there is no succinodehydrogenase activity. In this respect the type of metabolism is similar to that of schizonts and microgametocytes. The macrogamete leads an anaerobic existence. Energy losses during the growth and synthesis of a large number of granules are completely met by glycolytic respiration of the substrate which exists in sufficient quantities in the macrogamete and in the surrounding cytoplasm of the host epithelial cells.

In young macrogametes, alkaline phosphatase activity has been observed only in the karyosome of the nucleus. This has been found in *E. magna, E. intestinalis,* and *E. tenella* (Gill and Ray, 1954a; Beyer, 1960). In adult macrogametes of *E. magna,* alkaline phosphatase activity has been observed in the peripheral granules and particularly in the protein granules. In *E. intestinalis* these granules give no noticeable reaction to alkaline phosphatase.

Acid phosphatase has been observed both in the nucleus and cytoplasm of rabbit and chicken coccidia. In addition, activity has been discovered in glycoprotein granules and rather weakly in the cytoplasm. In the nucleus, the strongest acid phosphatase activity was in the karyosome and to a lesser degree in the nucleoplasm (Gill and Ray, 1954a; Beyer, 1960).

Electron microscope studies of macrogametes conducted by Scholtyseck (1963b) on *E. perforans* have established some interesting properties of their structure. The young macrogamete was covered with a membrane about 125 Å thick. In more mature macrogametes, numerous tubules about 650 Å in diameter became noticeable; the inside lumen of the tubules was about 400 Å in diameter and their length reached 1.3 microns. They were associated with the surface of the macrogamete and opened into the cytoplasm of the cell which surrounds the macrogamete. The membrane of the tubules had a transverse cross-hatching of lines, with dark stripes that were 90 Å across and light stripes 75 Å across. The width of the stripes was 165 Å. It may be assumed that the presence of such tubules aids in the penetration of fluid nutritive substances from the cytoplasm of the host cell into the cytoplasm of the gamete. This is a type of microfilament similar to those of the epithelium of the intestinal cells. Because of such microfilaments, the absorptive surface is greatly increased. Such tubules have not been observed with the electron microscope in macrogametes of *E. intestinalis* and *E. magna.*

A study of the macrogametes of *E. perforans* with the electron microscope has shown the presence of large mitochondria with short tubules in the cytoplasm. There are small bodies with internal membranes (the so-called H-bodies) which increase in size and number in young gamonts prior to the formation of glycogen. Apparently, these take part in the formation of the internal membranes of the oocyst and correspond to the protein granules described with light microscopy. Somewhat later, the dark bodies appear; it is possible that they correspond to the plastinoid-glycoprotein granules observed with light microscopy (Scholtyseck, 1963b).

## Fertilization and the Formation of Walls
## on the Surface of the Zygote

The process of fertilization has been poorly studied in representatives of the family Eimeriidae. This is because fertilization happens very quickly and is difficult to find in preparations. Schaudinn (1905) described the penetration of microgametes into the macrogametes of *E. schubergi* and Wedekind (1927) observed this process in *Barrouxia*. Among other representatives of the suborder Eimeriidea, fertilization has been observed in *Aggregata* (Dobell, 1925) and in the order Adeleida in *Klossia* (Naville, 1927). Apparently, other investigators have seen individual stages of fertilization, but the actual penetration of the microgamete into the macrogamete has never been recorded. Schaudinn (1900) noted for *E. schubergi* and Nieschulz (1922) for *E. labbeana* that several microgametes can penetrate into the macrogamete. One of these unites with the nucleus of the macrogamete, but the others are resorbed. According to Scholtyseck (1963a), in *E. maxima* only one microgamete penetrated the macrogamete.

The site where the microgametes penetrate has not been completely determined. Apparently, no special micropyle exists in the macrogamete, and the microgametes may penetrate at any point on the surface of the macrogamete. However, in *Pfeifferinella impudica* a special tube-shaped outgrowth has been described (Fig. 21); this is probably used for

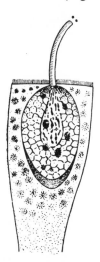

Fig. 21. Adult macrogamete of *Pfeifferinella impudica* with pipe-like micropyle, through which the microgamete passes (After Léger and Hollande, 1912).

penetration by the microgamete (Léger and Hollande, 1912).

The microgamete apparently penetrates into the macrogamete without the aid of flagella. Wedekind (1927) reported that in *Barrouxia* the flagella remain on the surface of the macrogamete. The same has been observed in *E. maxima* (Scholtyseck, 1963a). It is not known whether only the nucleus of the microgamete penetrates or whether the basal bodies and mitochondria also enter. With the aid of an electron microscope, no one has yet observed basal bodies or centrioles from microgametes inside of fertilized macrogametes.

The nuclei of the microgamete and the macrogamete approach each other, but no conjugation occurs at this point. Only during the first metagamic division of metaphase do the male and female chromosomes move into one equatorial plane. The first division begins after the wall has formed around the zygote. In *E. maxima*, Scholtyseck observed that even prior to the first division, the nucleus of the microgamete forms five easily seen chromosomes (Fig. 22). During the process of fertilization, the macrogamete nucleus undergoes only minor changes. In *E. magna* and *E. intestinalis*, nuclear volume decreases and a Feulgen-positive reaction is restored. In the fertilized macrogamete of *E. magna* and *E. intestinalis*, it is possible to observe the male nucleus alongside the female nucleus. Both take the Feulgen stain well.

After the microgamete penetrates the macrogamete, a delicate fertilization membrane is formed. At the same time, a substantial change occurs in the metabolism of the newly formed zygote. The activity of

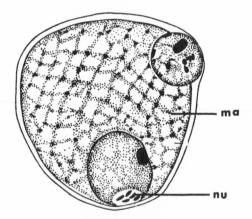

Fig. 22. Fertilization of *E. maxima* macrogamete (After Scholtyseck, 1953). *ma,* Macrogamete; *nu,* microgamete nucleus with five chromosomes.

succinodehydrogenase increases greatly as does the concentration of free and attached sulfhydryl groups. After fertilization, a change from anaerobic to aerobic existence occurs. At this same time, protective walls form around the zygote. This process has been studied in many coccidia with the aid of light microscopy (Wasielewski, 1904; Hadley and Amison, 1911; Reich, 1913; Henry, 1932b; Roudabush, 1937; Lapage, 1940, and others). Scholtyseck (1963b) used electron microscopy to describe formation of the oocyst wall of *E. perforans.*

Usually, the oocyst wall is formed from plastic elements which accumulate during the process of macrogamete growth. This process is similar among the various coccidia. Reich, Wasielewski, Hadley, Roudabush, and others noted that the macrogametes contain two types of granules (plastinoid and chromatinoid). These granules are distributed at the periphery of the macrogamete, and the walls are formed by fusion of the granules. Lapage (1940) described such a process in *E. caviae.* He stated that when stained with azan by Mallory's method, two types of granules are present. Some stain orange and others red. The red granules form the external oocyst wall by fusing together and the orange granules move to the periphery to form the internal wall.

In *E. magna* and *E. intestinalis,* large glycoprotein granules are situated in a solid layer along the periphery of the macrogamete directly beneath the external membrane. The granules reach a size of 2 microns. They then become oval and join together to form a solid layer. At first the layer is quite thick, but as the oocyst increases in volume the layer which becomes the external wall stretches and becomes thinner (Figs. 19 and 20).

Löser and Gönnert (1965) reported that in the macrogametes of *E. stiedae, E. falciformis,* and other species, the granules arising along the nucleus contained a mucous secretion and stretched out as the macrogamete grew until they reached the border of the cytoplasm and the host cell. They became arranged in a single layer. At this time the nucleus produced a serous secretion which stained an intense red-orange with azan. This secretion aided in sticking and pressing the peripheral granules of the mucous secretion into a homogeneous wall. A solid oocyst wall was formed by this process. In opposition to these data, Scholtyseck (1963b) suggested that there is a granular zone on the periphery of the macrogamete, the solidification of which leads to the formation of the walls. After the formation of both walls, the protein and glycoprotein granules remain in the cytoplasm of the oocyst and are apparently used for later construction of the sporocyst walls.

Electron microscope studies of the macrogamete after fertilization have enabled us to determine to some extent how the oocyst wall is formed. According to Scholtyseck (1963b), the unfertilized macrogamete of *E. perforans* is covered with a membrane which separates it from the cytoplasm of the cell. This is the so-called first external membrane. After the microgamete penetrates this membrane at the surface of the macrogamete, a second membrane (fertilization membrane) forms under the external membrane. This becomes the external membrane surrounding the outer wall of the oocyst. The sinuous membrane structures are visible in the granules forming the outer wall. A third internal membrane is also present. The formation of the internal wall is associated with the appearance of two more membranes, between which the material of the wall itself is situated. This is formed from the protein granules or from the "H" granules of the second type (named by Scholtyseck, 1964). Scholtyseck assumed that five membranes form around the zygote of *E. perforans* (Fig. 23). All of these are visible with the electron microscope. When examined with light microscopy, these membranes are not visible, and only the solid walls confining the membranes are visible (Fig. 23). Usually it is possible to see only two walls which are sometimes referred to as the ecto- and endocysts. According to Scholtyseck, the endocysts are not uniform in thickness.

Henry (1932b) studied the walls of various species of coccidia with the light microscope and concluded that not two, but three walls can be seen in some oocysts. She observed that in oocysts of *E. intricata* there is a thick (2 microns) and a rather unevenly stained wall inside of which is another thin (0.2 to 0.4 microns) and colorless wall. Both these walls lie against an internal, colorless wall. The delicate external surface of the wall is connected to the polar cap. The same was observed in *E. faurei*,

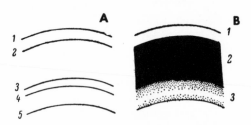

FIG. 23. Structure of the oocyst wall (After Scholtyseck, 1964). *A*, Diagram of the walls as demonstrated by electron microscopy; *1*, external membrane; *2*, external membrane of the surface wall; *3*, internal membrane of the external wall; *4* to *5*, external and internal membranes of the internal wall; *B*, walls visible with light microscopy; *1*, external membrane (sometimes not visible); *2*, external wall; *3*, internal wall.

*E. ninakohlyakimovae,* and *Isospora lacazei.* Henry reported that a third (most superficial) wall exists in many oocysts, but it is not always easy to distinguish it or separate it. However, in most cases the oocysts have two walls of varying thickness such as are observed in oocysts from rabbits and chicken coccidia. It is possible that the most superficial, delicate wall is a flattened membrane of the macrogamete which is easily destroyed and lost when the oocyst is freed from the epithelial cell and enters the lumen of the intestine. As a result, the oocysts may have only two walls. Observations of the oocysts of *Isospora felis* and *I. rivolta* indicate that these species have only one wall, which no one has been able to separate into layers by any known method (Kheysin, 1937a).

According to Monné and Hönig (1954), the internal wall of oocysts has a characteristic birefringence. This attests to the high organization of its submolecular structure. Apparently, the external wall also has the same properties. According to the same authors, the internal wall consists not only of proteins, but also contains lipids which form plates which solidly abut with one another. Other investigators (Pattillo and Becker, 1955; Tsunoda and Itikawa, 1955; Kheysin, 1958a, 1960) have failed to find lipids in oocyst walls. It is probable that the wall includes a lipoprotein complex which forms the structure of the basic membranes which cover the oocyst wall.

Several lipid-dissolving substances, such as alcohol [sic] easily penetrate oocysts. This is to some extent indirect evidence of the fact that lipids enter into the makeup of the internal wall of the oocyst. Some substances which do not dissolve lipids, such as heavy-metal salts, also have the ability to pass through the oocyst wall.

With electron microscopy, Holz (1954) observed plates containing glycogen in the surface of oocyst walls. However, glycogen has not been discovered in the wall by any microchemical methods. Scholtyseck and Weissenfels (1956) did not observe such plates in *E. tenella* although they could see the smooth surface of the oocyst.

It has not been determined whether acid mucopolysaccharides are present in oocyst walls as they sometimes are in the protective wall of the ova of parasitic nematodes and other parasitic worms. In the opinion of Ray and Gill, the oocyst wall of *E. tenella* contains acid mucopolysaccharides of the hyaluronic acid type. However, Pattillo and Becker (1955) and Kheysin (1958a, 1960) could not find this substance in oocyst walls. Apparently, only neutral mucopolysaccharides enter into the makeup of the walls.

Despite the fact that the oocyst walls give the same cytochemical reactions as the macrogamete granules from which they are formed, their physical properties are different. The walls are more resistant to certain chemicals than the granules. The walls cannot be digested by pepsin, but the granules swell and are partially destroyed. Under the influence of sodium hypochlorite, the outer wall breaks down, but the inner one remains unharmed. Monné and Hönig (1954) reported such resistance is due to the lipids which impregnate this protein wall. Sulfuric acid, which dissolves chitin or cellulose, has no effect on oocyst walls. This indicates a lack of such substances in the oocyst wall. Thioglycolate destroys the integrity of the walls, which is to some degree evidence of the presence of keratinoid substances in the wall.

Using Fontane's method, Monné and Hönig found that the external wall reduced silver nitrate in a solution of ammonia and stained brown or red. This method reveals phenols and quinone-tanned proteins. Thus, they concluded that the external wall contains quinone-tanned proteins, which are very solid scleroproteins. Apparently, this also explains the high resistance to various physical and chemical factors. Digestive juices do not break down these proteins. Löser and Gönnert (1965) reported that the oocyst wall contains o-dioxy-phenolprotein, which shows up when stained with an aqueous solution of 1% malachite green.

Consequently, the oocyst wall is very resistant, and as will be shown below, the internal wall is semipermeable and protects the zygote or sporozoites against the effects of various chemical substances, while the external one protects it mainly against mechanical damage. Most chemical substances penetrate the external wall.

### The Oocyst

The oocyst (zygocyst) formed after fertilization provides for the process of sporogony and protects the zygote and sporozoites from various unfavorable effects of the external environment. The oocyst (Fig. 24) is the stage in the cycle of coccidian development which leaves the host and can survive in the external environment for a given period of time. Of all the stages in the life cycle, this one lasts the longest. The oocyst makes it possible for a new host to become infected. This happens when the oocyst is ingested from the external environment.

Oocysts may be of different shapes. Round oocysts (spherical) are often found, e.g., *E. cylindrospora* from *Alburnus alburnus* and in many other species parasitic in fish (*E. carpelli, E. cotti, E. soufiae,* and others). Many representatives of the *Isospora* of birds, as well as *E. miyairii, E. falciformis, E. mitis, E. alpina,* and others have spherical oocysts.

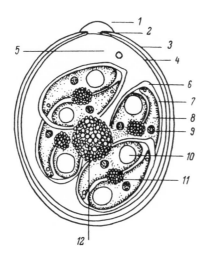

Fɪɢ. 24. Diagram of oocyst structure (After Levine, 1963). *1,* Polar cap; *2,* micropyle; *3,* external wall; *4,* internal wall; *5,* polar granule; *6,* Stieda body; *7,* sporocyst; *8,* sporozoite; *9,* sporozoite nucleus; *10,* sporozoite polar body; *11,* sporocyst residual body; *12,* oocyst residual body.

Subspherical oocysts are found in *E. parva, E. zurnii, E. subspherica, I. suis, E. debliecki,* and others. Oval, ellipsoid, and cylindrical oocysts are most often found, and transitional forms can always be found between them. Examples of such oocysts are *E. magna, E. irresidua, E. media, E. arloingi, E. intricata, E. ellipsoidalis, E. ninakohlyakimovae, E. brasiliensis, E. auburnensis, E. hagani, E. tenella.* Less often, pyriform and ovoid oocysts, such as *E. alabamensis, E. bukidnonensis, E. wyomingensis, E. anseris, E. intestinalis, E. piriformis,* and *E. faurei,* are found. Some of the oocysts are bottle-shaped *(E. roscoviensis, E. urnula),* flask-shaped *(Cyclospora argeatati),* urn-shaped *(E. granulosa),* kidney-shaped *(Isospora lieberkuehni),* or polygonal *(E. triangularis).*

The form of oocysts may vary to a considerable degree within a single species. This has been well shown in the literature on rabbit coccidia (Kheysin, 1940, 1947a, 1948, 1957a) and coccidia of chickens (Fish, 1931; Becker et al., 1955). The variability of oocyst shape within a species may depend on several factors which affect the development of the macrogametes in the host cells. The appearance of broadly oval and short oocysts in *E. intestinalis* is associated with a heavy infection in which several macrogametes (as many as three) develop within the same host cell at the same time. As a result each oocyst is not only smaller in size, but has more of a broad-oval form than the usual pear-shape. The variation in oocyst shapes of a rabbit coccidium are illustrated in Fig. 25.

A significant variety of oocysts is created by their varying wall thicknesses and the presence or absence of a micropyle and the cap which covers it. Oocyst walls may be of varying thickness. In some oocysts the external wall is significantly thicker than the internal one. For example, this is observed in *E. leuckarti, E. brinkmanni, E. picti, E. intricata,* and others. Sometimes it attains a thickness of 2 to 3 microns, while the internal wall does not exceed 1 micron. In *E. leuckarti,* the external wall may be 6.5 to 7 microns thick. Such oocysts have been observed by several authors to have a single-layered wall.

In many oocysts the external wall may be of the same thickness or thinner than the internal one. When this occurs both walls have a thickness of 2 to 2.5 microns, or the internal wall is 2 to 2.5 microns, but the external wall does not exceed 1 micron. It is sometimes considered that such oocysts have a double-layered wall. Such walls are found in *E. columbarum, E. phasiani, E. ellipsoidalis, E. alabamensis, E. arloingi, E. parva, E. faurei, E. granulosa,* and others. The oocysts of rabbit and chicken coccidia also have a more delicate external wall. The large oocysts of *E. noelleri* from the camel have a dense outer wall that is 10 to 15 microns thick, but the middle and internal walls are more delicate (Henry and Masson, 1932; Pellérdy, 1956).

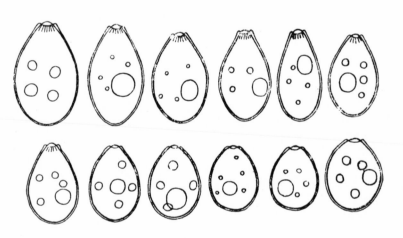

Fig. 25. Variability of the form and size of *Eimeria intestinalis* oocysts (After Kheysin, 1957a).

The external wall may be smooth or uneven and rough. The structure depends upon the uniformity of the material making up the external wall. Sometimes the external wall is composed of small or large granules *(E. sculpta, E. semisculpta, E. ponderosa, E. arloingi, E. auburnensis)*. There are oocysts with striped walls *(E. striata, E. fulva)*, wrinkled walls *(E. rugosa)*, rough walls *(E. scabra, E. intricata, E. superba, E. picti, E. maxima, E. callospermophili)*, or with spines *(E. spinosa)*. More often, however, the external wall is of a homogeneous nature and looks completely smooth and translucent; examples are the oocysts of rabbit, chicken, and turkey coccidia. Sometimes the walls contain a pigment, and the oocysts are colored. The external wall is often yellow or brown. The shade may be intense *(E. magna, E. irresidua, E. intricata, E. scabra, E. spinosa)* or light *(E. media, E. maxima, E. intestinalis)*. In *E. faurei* the wall is yellowish; the wall of *E. tenella* looks greenish. Wenyon (1926) described a reddish wall for *E. canis*. Some species have colorless walls *(E. zurnii, E. ellipsoidalis, E. mitis, E. acervulina, E. perforans)*. If the oocyst loses its external wall because of some sort of mechanical factor, it becomes colorless, since the internal wall is usually not pigmented (Kheysin, 1937a, 1947b). For the most part the internal wall is shaded light green or pink, depending on its optical properties. The thicker the external wall, the more intensely it is colored. A delicate wall sometimes appears colorless. The color of the wall is in some cases acquired by absorption of bile pigments in the intestine. The oocysts of the intestinal coccidia of rabbits taken from the cecum (e.g., *E. coecicola*) appear colorless, but the same oocysts taken from the feces are light brown.

All oocysts can be divided into two groups. Some have a micropyle and others do not. Round oocysts have no micropyle, but the oval, ellipsoid, and other shapes often have a micropyle at one end of the oocyst. The micropyle is a part of the oocyst wall which differs in structure from the rest of the wall. In some oocysts the wall at this point (always at one pole) is markedly thickened, but the inner wall is unchanged, e.g., *E. arloingi* and *E. perforans*. In other oocysts the external wall has a small, round aperture at the pole through which the internal wall protrudes slightly (Figs. 46 and 48). The edge of the external wall around such a micropyle is confined by a thickened cylinder so that the micropyle is clearly visible, e.g., *E. magna, E. irresidua, E. media, E. intestinalis*, and *E. canis*.

In some cases the micropyle appears as a thin spot on the internal wall, e.g., *E. granulosa*. However, it should be mentioned that the structure of the micropyle has been poorly studied, and at the present many details of its structure are unclear. In some pyriform oocysts, such as *E. roscoviensis,* one pole is stretched out in the form of a neck, and its free edge, which is bare of an internal wall, is covered by a delicate external wall. Labbé (1893) called this a pseudomicropyle, although it apparently corresponds to the micropyle of other oocysts. In those oocysts which lack a micropyle, both walls are sometimes slightly thinner at one pole (*E. maxima, E. tenella, E. faurei,* and others).

What is the functional significance of such a structure? Apparently, the sporozoites leave the oocyst through the micropyle after the oocyst enters the intestine of the host and is subjected to the digestive fluids. It is highly probable that the formation of a micropyle corresponds to that place in the macrogamete through which the microgamete enters.

In some species another small cap, which is a solidification of the external wall, forms above the micropyle (Fig. 46). Henry (1932b) reported that the cap formed as a thickening of the surface of the third wall which is sometimes present in oocysts of sheep coccidia. This cap is visible in *E. arloingi, E. crandallis, E. granulosa, E. intricata, E. punctata, E. alabamensis,* and *E. brasiliensis* [*E. alabamensis* does not have a cap. Ed.]. The cap apparently forms from material outside the fertilized macrogametes as the oocysts develop in the epithelial cell of the host. The cap can be easily removed from the surface of the oocyst, but such oocysts do not lose their protective properties. The presence or absence of a micropyle or cap has no effect on the resistance of the oocyst to damaging external influences. This is probably determined by the fact that the micropyle and cap are formed by the external wall, but the internal wall prevents outside material from entering the oocyst.

The sizes of oocysts vary to a considerable degree. Some species form very small oocysts (under 10 microns). Oocysts of *Cryptosporidium parvum* from the intestine of mice are 4.5 microns in diameter. The oocysts of *E. cheni* from the intestine of *Mylopharyngodon piceus* have a diameter of 8.5 to 9 microns. *Cryptosporidium meleagridis* from turkeys has oocysts that are 4.5 by 4.0 microns. Round oocysts with a diameter of 4.5 to 5.5 microns of *E. dogieli* are present in the kidneys of the freshwater fish *Opisthocentrus ocellatus.* Most oocysts are 15–30 microns by 10–25 microns. Most of the species parasitizing domestic animals have such oocysts.

Some oocysts have been found which attain very large dimensions. Their lengths vary from 35 to 100 microns and their widths from 25 to 80 microns. Such large oocysts are formed by *E. noelleri* from camels (80–100 by 60–80 microns), by *E. leuckarti* from horses (80–87 by 55–59 microns), by *E. gilruthi* from sheep (42–54 by 31–36 microns); and by *E. travassosi* from *Dasypus sexcinctus* (60 by 45 microns). *Barrouxia schneideri* from the intestine of *Lithobius impressus* has oocysts with a diameter of 40 to 80 microns. The oocysts of *E. sardinae* in the testes of sardines are 60 to 65 microns. These large oocysts have a very thick external wall which is dark brown.

The largest oocysts are formed in those species in which the macrogametes develop in the connective tissue (e.g., *E. gilruthi* and *E. leuckarti*), where there is practically no limit to the size they might attain. The oocysts of *E. noelleri* develop in epithelial cells embedded in the connective tissue of the mucosa of the upper small intestine (Pellérdy, 1965). The smallest oocysts are formed in those species which develop outside the cells on the surface of the intestinal epithelium, e.g., *Cryptosporidium parvum*.

Oocyst size sometimes varies considerably within a single species. This has been noted for many species of rabbit coccidia, as well as those of chickens, sheep, and other animals (Boughton, 1930; Fish, 1931a and b; Jones, 1932; Kheysin, 1940, 1947a, 1948, 1957a; Becker et al., 1955, 1956; Krylov, 1960; Kogan, 1962, 1965). Differences in the sizes of oocysts of one species, observed in infections of the host by a single oocyst, have in some cases been so great that the investigators began to suspect the presence of several species of parasites if the source of the original inoculum was in any way doubted. Variability in oocyst size depends on a series of factors which determine the development of macrogametes. It has been observed that the average length and width of oocysts of *E. magna* and *E. intestinalis* from rabbits and *E. tenella* and *E. acervulina* from chickens decreased when the inoculation dosage was increased (Jones, 1932; Kheysin, 1947a, 1957a). This is so, because several macrogametes can develop in each host cell, but the ultimate size each will attain is severely limited by the size of the host cell. As a result, the gametes accumulate a small amount of nutritive material and become fertilized without having reached full size, after which they yield undersized oocysts. This phenomenon has not been observed in *E. perforans* and *E. necatrix*.

At the end of the patent period of rabbit coccidia smaller oocysts appear than at the beginning of the patent period; this is probably associated with a certain retardation of macrogamete growth at the end of the patent period when the rabbit is developing a protective reaction. No such phenomenon has been noted for coccidia of birds (Becker et al., 1955, 1956). The dimensions of the oocysts may vary greatly during the development of the endogenous stages in various individuals of one host species. Apparently, some of the host's individual properties have an effect on oocyst formation.

Kogan (1962, 1965) reported that the host's diet may influence the sizes of the oocysts of *E. necatrix*. In chickens on a protein diet, larger oocysts were produced than in those on a grain diet. The size of the oocysts on the protein diet increased by 10 to 15%. The author thought this phenomenon was due to a high level of nourishment of the intestinal epithelial cells which grow correspondingly and permit the parasites to grow larger.

Reyer (1937) observed that the size of the oocysts of *Barrouxia schneideri* from *Lithobius* depended on the temperature at which the centipedes were kept. If kept at a low temperature, the animals yielded much larger oocysts than those obtained from centipedes which were kept at high temperatures, but the dimensions of the sporocysts remained unchanged. He also found that larger oocysts are produced by hungry *Lithobius* individuals than by well fed ones. An analogous phenomenon was observed by Zaika and Kheysin (1959) in oocysts of *E. carpelli* from carp. At higher temperatures the fish yielded smaller oocysts, but at low temperatures larger oocysts were formed.

## SPOROGONY

### The Formation of Sporocysts and Sporozoites

The oocysts formed in the cells of the intestinal mucosa or other organs (e.g., nephrons, bile ducts of the liver, seminal vesicles, etc.) are ultimately eliminated from the host into the external environment. There, development is possible only in aerobic surroundings. However, the oocysts of some species of coccidia have the ability to sporulate while in the tissues of the host, i.e., under conditions of considerable oxygen deficit. Such oocysts are eliminated from the host in a sporulated state. Examples are *E. carpelli* and *E. subepithelialis* (from fishes), *I. bigemina* (from dogs and cats), and *I. dirumpens* (from snakes).

The oocysts of most coccidian species are passed into the external environment in an unsporulated state, and in such oocysts which have left the host cells and entered, for example, the lumen of the intestine, the entire energy regimen changes when compared with that of the previous location of the macrogametes and zygotes. It has been found that in oocysts of *E. intestinalis* and *E. magna,* succinodehydrogenase activity was present, and the concentration of both free and protein-connected sulfhydryl groups increased (Beyer, 1963b). The oocysts were in an environment rich in oxygen, and aerobic oxidation of the substrate proved more economical. The large supplies of glycogen and fat accumulated by the macrogamete are used in the oocyst for the process of sporogony and to maintain the viability of the sporozoites.

When the oocysts emerge from the epithelial cells or connective tissues into the lumen of the intestinal wall, the cytoplasm of the zygote usually fills the entire oocyst and clings tightly to the internal oocyst. By the time the oocysts emerge with the feces into the external environment, the cytoplasm of the zygote has begun to shrink away from the wall and forms a sphere. The formation of the spherical mass of the zygote is associated with dehydration. The fluid which is excreted from the cytoplasm of the zygote remains in the space between the zygote and the oocyst wall. The presence of fluid in the oocyst is demonstrated by the fact that the small, light-refracting bodies lying within the oocyst undergo a characteristic Brownian movement. Bacteria have sometimes been observed in this fluid (Yakimov and Timofeyev, 1940; Kheysin, 1947b). The sporozoites which emerge from the sporocysts because of the influence of enzymes also move within the oocyst; this would be possible only in a fluid.

In all coccidia the entire process of sporogony can be observed only in living forms (never in stained preparations). This is because stains cannot penetrate the oocyst wall, and because oocysts cannot be made into permanent preparations. Inasmuch as the oocyst walls can be penetrated by fixatives and stains, it is possible to observe several stages of sporogony in fixed and stained preparations of oocysts which sporulate without leaving the host's body.

The process of oocyst sporogony in the external environment generally proceeds according to a unified plan that has been observed in many species of coccidia. It is always possible to see a light, bubble-like nucleus in the center of the spherical zygote, around which numerous light-refracting granules are concentrated. In various species of rabbit and chicken coccidia, it has been observed that protein granules are located near the periphery.

The division of the oocyst nucleus of most coccidia, in which sporogony occurs in the external environment, is difficult to observe. It is possible to find nuclear division, but only with phase-contrast microscopy; however, the details of this process are impossible to determine. Externally, the division of the nucleus is accompanied by several changes in the cytoplasm. For example, this process in *E. magna* (Fig.

26) will be described below. The first feature of zygote division is expressed as a concentration of lipid granules around the nucleus. Later, the granules are passed into the residual body of the oocyst. In those species which do not have an oocyst residuum, the lipid granules do not concentrate at the beginning of division. At this time, the nucleus of the zygote divides twice to form four nuclei. As soon as nuclear division is complete, four protrusions, directed in mutually perpendicular directions, appear on the spherical surface of the zygote. The protrusions gradually increase in size and become prominent. Light colored vacuoles are visible on the free ends of the protrusions. The protrusions increase in size, become spherical, and finally separate from one another to form spherical sporoblasts. The residual body always lies between the sporoblasts of *E. magna* and other coccidia which have a residual body. The residual bodies differ from sporoblasts in their charcteristic light refracting properties. Further changes in the sporoblasts are as follows. One end of the spherical sporoblast begins to elongate, and the opposite side becomes wider and flatter. Finally, the sporoblasts assume a pyriform or pyramidal shape. This stage is often called the pyramid stage. The four pyramids are joined at their bases in the center along the sides

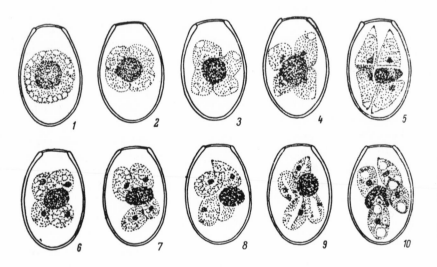

Fig. 26. Sporogony of *E. magna* oocysts (After Kheysin, 1947b). *1*, Spherical zygote; *2* to *4*, beginning of the formation of the four sporoblasts; *5*, pyramid stage; *6* and *7*, end of sporoblast formation; *8* and *9*, formation of the sporocysts; *10*, formation of the sporocysts and sporozoites.

of the residual body (if it exists such as in *E. magna*). The peaks of the pyramids are then pointed in various directions toward the oocyst wall. A bright vacuole and a dense, light-refracting body are always visible at the peaks of the pyramids. Glycogen and lipid granules concentrate at the base of the pyramid. Later, these enter into the structure of the sporocyst residuum.

The pyramidal stage is rather short-lived, as compared with the entire process of sporogony. Later, the pyramidal sporoblasts again become rounded, and each spherical sporoblast contains many vacuoles and a granular mass which forms the future residual body. The spherical sporoblasts then begin to stretch and become oval-shaped. At this time protein granules are situated along the periphery of the sporoblast. Later, it is possible to observe a delicate wall develop around the oval sporoblasts and the sporoblast develops into a sporocyst.

A single division of the nucleus occurs in the sporocyst, after which the sporocyst cytoplasm divides into two longitudinal parts; the result is the formation of two sporozoites. At the same time, the sporocyst sheds its residual body, into which go light-refracting lipid and glycogen granules. After the sporozoites form in the sporocyst, it is possible to speak of a sporocyst. In other species of this genus [*Eimeria*. Ed.], sporogony proceeds in an analogous manner.

The entire process of sporogony in *Eimeria* can be divided into three phases. The first period is composed of two parts during which there is a division of the zygote nucleus followed by the preparation of the cytoplasm for further rearrangement. The second is one of a cytoplasmic rearrangement of the sporoblasts during which the pyramidal stage and oval sporoblasts are formed, leading to the formation of sporocysts. During this period the nucleus does not undergo division. The third period is the formation of sporozoites within the sporocysts, accompanied by synchronous division of the sporocyst nuclei. The second period, during which differentiation of the sporoblasts and sporocysts occurs without nuclear division, lasts the longest.

Because two sporocysts, each containing four sporozoites, are formed in *Isospora*, sporogony differs somewhat from that which occurs in *Eimeria*. In *Isospora*, sporogony occurs in a manner similar to that of other genera, so the description here will be based on Schwalbach's (1959) data for *I. turdi, I. sylviae*, and other *Isospora* species of various birds (Fig. 27). Sporogony, as in *Eimeria*, is begun by a shrinking of the zygote into a sphere. Often one or several small, light-refracting bodies, which remain within the oocyst wall until the end of sporogony, are eliminated during this phase. Similar bodies are sometimes eliminated in

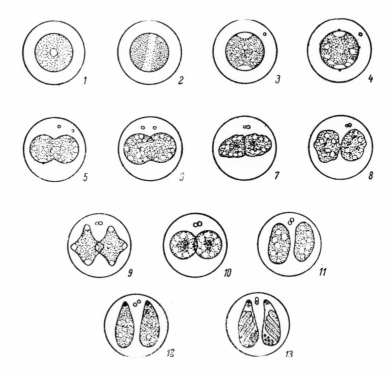

FIG. 27. Sporogony of *Isospora* oocysts from birds (After Schwalbach, 1959). *1*, Spherical zygote; *2*, first nuclear division; *3*, nuclei at the poles of the zygote; *4*, formation of the polar granules; *5*, division of the nuclei and start of the zygote division into two sporoblasts; *6*, two sporoblasts; *7* to *9*, formation of the tetrahedron stage and division of the nuclei; *10*, formation of the spherical sporoblasts; *11*, stretching of sporoblasts; *12*, transition to sporocysts; *13*, adult sporocysts with sporozoites.

oocysts of *Eimeria* (e.g., *E. necatrix, E. mitis, E. tenella, E. brasiliensis*).

The first evidence of division in oocysts appears as a bright band which transects the spherical zygote. This band corresponds in location to the spindle of the dividing synkaryon. The two nuclei thus formed become located at opposite poles of the cytoplasmic sphere and are visible as bright spots in the granular cytoplasm. Small, light-refracting granules and numerous vacuoles appear on the surface of the cytoplasm near the nuclei. The second division of the nuclei begins before the zygote divides into two sporoblasts. As soon as the nuclei divide, a divisional fissure appears around the meridian of the cytoplasmic sphere.

The fissure gradually deepens, and finally the spherical zygote divides into two sporoblasts, each of which contains two nuclei. There is a light-refracting granule next to each nucleus which is analogous to the granules which are visible at the peak of the pyramid of the sporogony phase of *Eimeria*.

Four short pyramidal outgrowths, which correspond to the four nuclei, form on the surface of the sporoblasts. This stage resembles the pyramidal stage of sporogony of *Eimeria* oocysts. These sporoblasts, with the four outgrowths, usually develop into tetrahedron stages which are expressed in differing degrees among the various species of *Isospora*; they are especially prominent in *I. sylviae.* This stage corresponds to a subsequent division of each sporoblast. As a result, four nuclei are formed in each; these are situated so that they lie across from one another. The tetrahedron stage develops rather quickly into another spherical stage. The sporoblasts then become oval, and a wall is formed around each to form the sporocyst. Four sporozoites and a large residual body containing glycogen and lipids appear within each sporocyst. In some species of *Isospora,* a micropyle, with a small cap covering it, appears at one end of the sporocyst at the time the sporozoites are formed.

Thus, sporogony of the oocysts of *Isospora* differs from that of *Eimeria* by an alternation of the periods of cytoplasmic differentiation and nuclear division. The first period of double nuclear division in *Isospora* and *Eimeria* precedes division of the zygote into sporoblasts. Later, during the second period, a long phase of cytoplasmic differentiation with formation of sporoblasts, pyramidal stages, and sporocysts (without nuclear division) occurs in *Eimeria*. During this same period in *Isospora,* the zygote divides into two sporoblasts, and a tetrahedron stage is formed when the third nuclear division occurs. During the last period of sporogony of *Eimeria,* the nuclei of the sporocysts undergo a third division and form sporozoites. In *Isospora,* the formation of sporocysts and sporozoites is brought about by differentiation without nuclear division, which takes place much earlier.

Sporogony and oocysts of *Caryospora argentati* were studied by Schwalbach (1959). In the spherical zygote, the nuclei undergo a triple division, and eight sporozoites develop from stick-shaped outgrowths on the surface of a residual body. Such periodicity, as is observed in *Eimeria* and *Isospora,* cannot be found during sporogony of *Caryospora.*

Sporogony of *Wenyonella* is similar to that of *Eimeria*. The differences are that the nuclei of the sporocyst of *Wenyonella* divide twice, but in *Eimeria* they divide only once. As a result *Wenyonella* has four sporozoites, but *Eimeria* has two (Hoare, 1933). In *Barrouxia schneideri* the nucleus of the zygote undergoes multiple and repeated division. The nuclei distribute themselves along the periphery of the zygote cytoplasm, and then the sporoblasts, each containing a nucleus, become separated. A wall develops around each sporoblast to form a sporocyst which contains a single sporozoite.

### Sporocysts

The sporocysts which are formed within the oocysts are usually spherical, oval, ellipsoidal, pyriform, ovoid, or spindle-shaped. In most cases one end is tapered and the other broadened. Sometimes the sporocysts are lemon-shaped (e.g., *Barrouxia schneideri*), an elongate hexagon *(E. anguillae, E. alburni)*, a cylinder *(E. cylindrospora)*, or a bottle *(Isospora ampullacea)*. The length and width of the sporocysts are approximately one-fourth to one-half the corresponding dimensions of the oocysts. It is difficult to find any correlation between the sizes of the sporocysts and oocysts. In large oocysts the sporocysts are often rather large. The sporocysts are sometimes situated in the oocysts so that their longitudinal axes are parallel to the longitudinal axes of the oocyst. This is especially noticeable in long and narrow oocysts. However, the sporocysts may be arranged in a random manner, which is their usual position in round or spherical oocysts.

The oocysts are covered by a double wall which is not penetrable by most chemical substances. However, in many coccidia the wall of the sporocysts can easily be penetrated by several substances which cannot pass through the oocyst wall. Many stains cannot pass through the oocyst wall but do pass through the sporocyst wall. Sporocysts removed from oocysts are easily stained with vital dyes. Potassium iodide readily passes through the sporocyst wall and stains glycogen. As a rule, oocysts with solid walls have sporocysts with delicate walls. This is especially true of the coccidia of chickens, rabbits, cattle, sheep, and other land animals. The *Eimeria* of fish and amphibians or many *Isospora* of land animals have rather delicate oocyst walls which easily admit water, but the sporocyst walls appear very solid and apparently play a large role in protecting the sporozoites from unfavorable environmental conditions. For example, in *Isospora bigemina* the oocyst wall is sometimes destroyed and individual sporocysts are passed into the external environment with the feces.

Many species have a dense, light-refracting body, which is called a

Stieda or Schneider body, at the pointed and narrowed end of the sporocyst. This is a plug covering the micropyle of the sporocyst wall. The sporozoites pass through this micropyle.

During cytochemical investigations of the sporocyst wall of *E. acervulina,* Pattillo and Becker (1955) found that protein and polysaccharides were present (Fig. 28). In *E. magna* the wall of the sporocyst reacts only for protein (Kheysin, 1960). If the wall of the sporocyst gives a positive periodic acid-Schiff reaction, the Stieda body remains unstained and contains only the basic proteins. Granules under the wall also give a positive PAS reaction for proteins. These reactions do not disappear after treatment with diastase and pepsin.

Schwalbach (1959) studied the development of the sporocysts of various species of avian *Isospora* in detail and reported that the micropyle

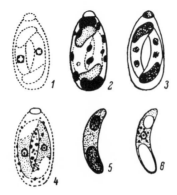

FIG. 28. Sporocysts and sporozoites of *E. acervulina* (After Pattillo and Becker, 1955). *1,* Sporocyst stained by the Feulgen method; *2,* Sudan red stain; *3,* Sulema-bromphenol stain for proteins (blue); *4,* toluidine blue stain for ribonucleic acid (metachromatic granules visible); *5,* sporozoite stained for proteins; *6,* sporozoite stained for ribonucleic acid.

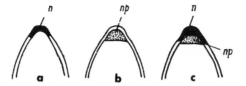

FIG. 29. Structure of the Stieda body of sporocysts of *Isospora* from birds (After Schwalbach, 1959). *a,* Sporocyst with button (*n*), micropyle; *b,* sporocyst with plug (*np*), deuteropyle; *c,* sporocyst with button and plug, macropyle.

has a varied and relatively simple structure (Fig. 29). The micropyle of the different species of *Isospora* consists of a solid cap or button which is a dense part of the sporocyst wall, but it is formed from a less dense material than that forming the plug lying under the cap. In some species there is only a cap or button on the tapered or pointed end of the sporocysts (e.g., *I. sylvianthina, I. lickfeldi, I. ampullacea, I. wurmbachi, I. lacazei*). This group is known as the Mikropylia (Fig. 29). Another group of species has only a plug in the micropyle, but does not have a cap. These are the Deuteropylia (Fig. 29). *Isospora turdi, I. phoenicuri,* and *I. sylviae* are assigned to this group. A third group has both a cap and a plug. These are the Makropylia (Fig. 29). This type of micropyle structure is found in *I. ficedulae, I. anthi, I. dilatata,* and *I. hirundinis.* The micropyles of the various *Eimeria* sporocysts have the same basic structure.

The number of sporocysts within the oocyst may vary. In representatives of the family Eimeriidae this characteristic is used as a criterion for division of the families into subfamilies. The subfamily Caryosporinae forms only one sporocyst. Two genera (*Mantonella* and *Caryospora*) are representatives of this subfamily. The subfamily Cyclosporinae contains coccidia in which two sporocysts are present in each oocyst. These are the disporids and include the genera *Cyclospora, Isospora,* and *Dorisiella.* The subfamily Eimeriinae includes members which have oocysts with four sporocysts, e.g., *Eimeria, Wenyonella,* and *Angeiocystis.* The subfamily Yakimovellinae is characterized by eight sporocysts and includes the genera *Yakimovella* and *Octosporella.* Recently, oocysts containing 16 sporocysts were observed in the subfamily Pythonellinae, which contains the genera *Pythonella* and *Hoarella.* Finally, an undetermined number of sporocysts apparently forms in oocysts of *Barrouxia* (subfamily Barrouxinae). Representatives of the family Aggregatidae have oocysts containing numerous sporocysts, the number of which apparently exceeds 32. A like number of sporocysts are formed in coccidia whose systematic position is not yet clear (*Merocystis, Caryotropha, Myriospora, Ovivora,* and *Pseudoklossia*). Thus, the number of sporocysts is a permanent one for a definite systematic group of coccidia (Hoare, 1933; Peraza, 1963).

### The Residual Body of the Oocyst and Sporocyst and the Light-Refracting Bodies

Residual bodies form in the oocysts of many species of the genus *Eimeria* during sporulation. These are clusters of lipid granules eliminated from the cytoplasm of the zygote during sporogony. The oocyst residuum is a characteristic feature of the oocysts of many species of coccidia. For example, there is a residual body, the size of which varies from species to species, in the rabbit coccidia *E. magna, E. media, E. coecicola, E. perforans,* and *E. intestinalis.* Residual bodies have never been reported in oocysts of *E. piriformis* and *E. irresidua.* It is interesting that the oocysts of all species of coccidia of chickens, cattle, and sheep have no residual body, nor is there a residual body in oocysts of *Isospora.* However, among the numerous species of rodent coccidia there are some which have residual bodies (Musayev and Veysov, 1965).

It is difficult to determine why some species of coccidia have residual bodies and others do not. Apparently, the nature of the metabolism of the macrogametes determines the accumulation of lipids. If many lipids accumulate, a considerable amount will not be used during sporogony and will form a residual body. The residual bodies of rabbit coccidia vary from 2.4 to 12.5 microns. The largest residual body is found in *E. magna* and has an average diameter of 6.7 to 9.6 microns. In *E. intestinalis,* the diameter of the residual body is 5.7 to 7.3 microns, in *E. media* 4.6 to 6.1, in *E. coecicola* 3.9 to 5.9 and in *E. perforans* 2.4 to 4.3 microns. In heavy infections, the dimensions of the residual body also decrease as compared with those from light infections. For example, in *E. magna* the diameter of the residual body is 8.1 to 9.7 microns. At the end of the patent period the diameter of the residual body may diminish. This is due to the lack of accumulated lipids in the macrogametes which develop near the end of the cycle. For example, in *E. magna* on the 1st day of the patent period, the diameter of the residual body averages 9.0 microns, but it is 6.7 microns at the end; in *E. media* it is 5.8 to 4.7 microns and in *E. coecicola* 5.3 to 4.0 microns, respectively. No external influences have ever been able to cause the appearance of a residual body in any species of coccidium which did not normally have them in its oocysts (Kheysin, 1947b).

In addition to the residual body of the oocyst, and perhaps totally unrelated to it, another small, light-refracting body or group of bodies appears in the oocyst during sporulation. These are either situated between the sporoblasts or lie near the oocyst wall such as in coccidia of chickens. In some species (e.g., *I. turdi*) the light-refracting bodies appear before sporulation begins. The refractile bodies are about 3 to 4 microns. These bodies do not dissolve in water, weak solutions of sulfuric, acetic, or hydrochloric acid nor in bases, but they partially dissolve in concentrated sulfuric acid. In *Isospora* these bodies push their way out of the cytoplasm of the zygote. Schwalbach was of the opinion that they represented some kind of excretory product formed at the start of sporulation.

An internal residual body may form within the sporocysts. These are spherical or sometimes consist of numerous irregularly distributed light-refracting granules. The diameter varies from 1–2 by 8–12 microns. Their composition is similar to that of the oocyst residuum and they are formed from large lipid droplets. Glycogen is distributed between the lipid droplets (glycogen is present in oocyst residua).

The oocyst residuum persists for a long time in oocysts which are found in the external environment; thus it is apparently not used as an energy source for maintaining the viability of the sporozoites. Over a period of 2 years there is an insignificant decrease in the sizes of oocyst residua of *E. intestinalis* and *E. magna*. The original diameters of the oocyst residua of *E. intestinalis* averaged 6.25 microns, and after 21 months at a temperature of 20° C in dichromate, the size was 5.9 microns. In *E. magna* the initial diameter of the residual body was 10.2 microns, but after 20 months of aerobic conditions the diameter was 9.7 microns. The oocyst residua underwent practically no decreases in size, and the material apparently was not used by the sporozoites.

Such is not the case with the sporocyst residuum. Pérard (1924) hypothesized that the sporocyst residuum of *E. perforans* was nutritive material which was used for maintenance of sporozoite viability. He noted that the sporocyst residuum gradually decreased in size when the oocysts were kept in the external environment. Experiments designed to study this process convincingly indicated that the sporocyst residuum of various species of coccidia changed when the oocysts were in the external environment (Kheysin, 1959a). The sporocyst residuum of *E. intestinalis* changed considerably after 20 months in dichromate at 20 C. The original diameter was 5.78 microns but at the end of observation the size was only 1.5 microns. After 3 to 4 months the sporocyst residuum lost the ability to take an iodine stain, indicating the disappearance of glycogen. At this point the fat droplets became larger and more scattered. Within 30 months the residuum completely disappeared or only individual fat droplets remained between the sporocysts. The 16- to 18-month old oocysts, from which the sporocyst residuum had practically disappeared, lost their infectivity; this was determined by experimental inoculation. Rabbits were only slightly infected by such oocysts despite the fact that they were given lethal doses. The same was observed in *E. magna*. In freshly isolated oocysts the residual body had an average diameter of 5.8 microns, but after 20 months the diameter was only 2.9 microns.

In *E. stiedae* the oocyst residuum consisted of three or four combined fat droplets. When stored for a long time under aerobic conditions, the residual body collapsed, but the granules remained. At the same time, the large sporocyst residual body (diameter up to 9 microns) decreased to 4 microns after 19 months. In *E. irresidua* an oocyst residuum was absent, but a sporocyst residuum as large as 12 microns was present. Within 20 months the sporocyst residuum was 5 to 7 microns and could no longer be stained with iodine. This demonstrated that the glycogen part of the sporocyst residuum is used before the lipid part.

The disappearance of the sporocyst residuum is clearly noticeable in oocysts of *Isospora*. The large sporocyst residuum of *I. rastegaievi* from the hedgehog had a diameter of 5 to 8 microns; in *I. felis* and *I. vulpis* from foxes the dimensions were 12 to 15 microns by 9 to 11 microns. In *I. laidlawi* from minks the sporocyst residuum was 16 by 13 microns. Within about 10 to 15 days after sporulation in the external environment, the residual body becomes friable and stains poorly in iodine; the fat droplets enlarge and disperse. Within 30 days, one or two large fat droplets are all that remain of the residual body of *I. vulpis*. In *I. felis* this occurs in 25 to 40 days. Complete collapse of the sporocyst residuum of *I. vulpis* and *I. laidlawi* is observed within 40 or 55 days. In *I. rastegaievi* the residual body persists as long as 5 months. After the oocysts have been in the external environment, the residual body disappears. These changes in the residual body have been observed in both aerobic and anaerobic surroundings (Kheysin, 1959a). Thus, the residual body of the sporocyst actually contains reserve food substances which are used by the sporozoites while in the external environment.

### Sporozoites

Sporozoites are the final stage of sporogony. They can undergo further development only if the oocyst is ingested by a host. There are three stages in the life of sporozoites. The first is a state of rest within the oocyst in the external environment. The second is the stage of active excystation after which the sporocyst enters the digestive tract of the host, and the third stage is that of existence in the intestine with subsequent penetration into the epithelial cells of the intestine, as well as the liver and other organs of the host. The structure of the sporozoites to one degree or another provides for accomplishment of these stages in the life of coccidia.

Sporozoites are pyriform, cylindrical, vermiform or pin-shaped (Figs. 28 and 56). Sometimes one end is pointed and the other is broad and rounded. Usually a large, light-refracting body, which is of a protein nature, lies in the wide part of the sporozoite. During sporulation, the refractile body is formed at the very last moment of sporozoite differentiation. After sporozoites of *E. acervulina* and *E. brunetti* were placed in pepsin for 1 hour, the light-refracting granules continued to stain in protein-sensitive colors just as intensively as prior to the application of pepsin (Pattillo and Becker, 1955). Hence, the body contains a large concentration of protein. The refractile bodies gave a weak reaction for lipids and a negative reaction for polysaccharides.

The center of the sporozoite usually contains a bubble-shaped nucleus with a peripheral distribution of chromatin. There is a small amount of RNA in the form of basophilic granules and a considerable amount of glycogen in the cytoplasm of the sporozoites of *E. acervulina* and *E. brunetti* (Fig. 28). The sporozoites of *E. magna* and *E. intestinalis* stain deep brown in Lugol's solution. The refractile bodies will not stain with iodine. Glycogen has also been observed in the sporozoites of *E. falciformis, E. stiedae,* and *E. tenella* (Giovannola, 1934; Edgar et al., 1944). Gill and Ray (1954b) did not observe glycogen in *E. tenella* because they were studying sporozoites which had penetrated into epithelial cells. It is probable that such sporozoites used all their glycogen while in the lumen of the intestine. The refractile body occupies approximately one-fourth of the length of the vermiform sporozoites of *E. carpelli* and *E. subepithelialis.* These long, vermiform sporozoites lie in the sporocysts with a bend in the middle of the body. The pyriform or pin-shaped sporozoites are almost always situated in the sporocysts with the pointed end oriented in different directions. In rare cases they lie with their narrow ends in one direction.

Vermiform sporozoites are most often found in oocysts of *Isospora,* and the pyriform and pin-shaped forms occur in *Eimeria.* The vermiform sporozoites are approximately 22 to 25 microns long and 1.5 to 3 microns wide. Sporozoites with other shapes are somewhat shorter and average 10 to 15 microns, but their width is about the same as that of the vermiform ones.

It is very difficult to observe the slow movements of sporozoites. The long, vermiform sporozoites of *E. carpelli* bend and contract. Sometimes a wave of contraction passes along the body, and the sporozoites move forward (Kheysin, 1947b). In some cases such long sporozoites move like a snake. Schwalbach (1959) observed the bending and snake-like movements of sporozoites in the sporocysts of *Isospora hirundinis.* During excystation, the sporozoites of *E. acervulina, E. bovis,* and *E. tenella* actively bend while passing through the micropyle; they twist and contract slightly during this time (Ikeda, 1956b; Doran and Farr, 1961; Doran, Jahn, and Rinaldi, 1962; Nyberg and Hammond, 1964). Although no contractile elements have been observed in sporozoites, they have not yet been studied with the electron microscope. With light microscopy, *E. tenella* sporozoites stained with hematoxylin and eosin azure (Tyzzer, 1929), and live sporozoites of *E. schubergi* and *Barrouxia* showed a longitudinal striping which may correspond to contractile fibril elements; these observations need further confirmation.

There are many mitochondria in the sporozoites. With light microscopy, the sporozoites of *Legerella testicula* have been shown to contain osmophilic granules which may correspond to elements of the Golgi apparatus (Tuzet and Bessiére, 1945).

The number of sporozoites formed in the oocysts may vary. The lowest number is four (*Cryptosporidium, Mantonella,* and *Cyclospora*). Sporozoites form from the zygote as a result of a double division of the nucleus. In representatives of the first species they lie directly within the oocyst; in the second, in one sporocyst and in the third, two are in each sporocyst (Figs. 30 and 31). In representatives of other species the number of sporozoites may increase to 8, 16, 32, 64, or more. Many sporozoites form in the oocysts of *Lankesterella, Barrouxia, Merocystis, Myriospora, Aggregata, Angeiocystis,* and others. Apparently, their number can reach 124 or 248 or perhaps even more.

In the oocysts of *Tyzzeria, Pfeifferinella,* and *Schellackia,* eight sporozoites lie free in the oocyst. In *Caryospora* all are in one sporocyst. In *Isospora* they are in two sporocysts, and in *Eimeria* they are in four sporocysts. Sixteen sporozoites are either in two sporocysts *(Dorisiella)* or in four sporocysts *(Wenyonella).* Thirty-two sporozoites may occur as two each in 16 sporocysts *(Hoarella).* In *Pythonella* each of 16 sporocysts contains four sporozoites.

The number of sporozoites depends on the number of metagamic divisions of the zygote nucleus. In representatives of the family Eimeriidae, the smallest possible number is two, and the largest is six, leading to the formation of 64 sporozoites. In the oocysts of *Yakimovella* no less than 64 or 120 sporozoites seem to develop (these are not really accurate observations). There are probably six or seven metagamic divisions in these oocysts.

An irregular number of sporocysts and sporozoites forms in *Barrouxia* oocysts. In *B. alpina* six to eight sporocysts are formed; in *B. schaudinni,* from four to 30 (more often 10–16) form; in *B. bulini* there are 18 to 42; in *B. belostomatis* 12 to 25; and in *B. caudata* 20 to 80. Thus, in this genus the number of sporocysts, and correspondingly spo-

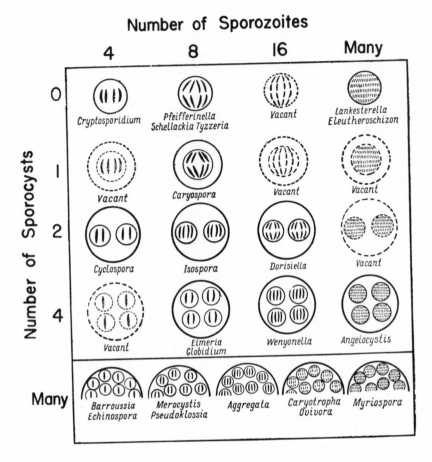

FIG. 30. "Periodic system" of the homologic series of coccidia in relation to the number of sporocyst and sporozoites (After Hoare, 1933).

rozoites, is not definite as is characteristic of other genera. In the family Aggregatidae the number of divisions may vary and probably reaches seven or eight, since the oocysts always contain a large number of sporozoites.

The numbers of sporozoites formed in the oocyst and the numbers in each sporocyst are now used as a generic criterion for each subfamily in the family Eimeriidae (Hoare, 1933, 1956; Peraza, 1963). Hoare (1933) observed that in each subfamily a parallel is found in the generic features. After comparing the number of sporocysts and the number of

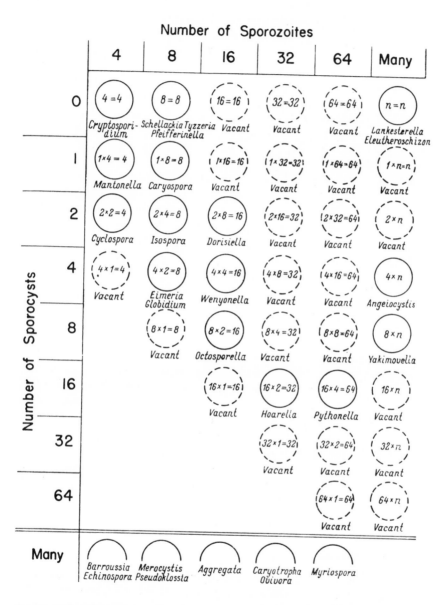

FIG. 31. "Periodic system" of the homologic series of Coccidia (After Peraza, 1963).

sporozoites, Hoare compiled an original table representing a unique periodic system of classification for the Eimeriidae (Figs. 30 and 31). The vertical rows of genera differ from one another in the total number of sporozoites in the oocysts (4, 8, 16, 32, 64), with each member (genus) in this vertical group differing from the others in the same vertical row by the number of sporocysts in which the sporozoites are distributed. The number of sporocysts in each vertical row changes from 0 to 1, 2, 4, 8, etc. The horizontal rows differ from each other in the number of sporocysts within the oocysts (no sporocysts, 1, 2, 4, or 8 sporocysts). In each such row, the genera differ from one another by the number of sporozoites in the oocyst and sporocysts. If the minimum number is four, then each succeeding genus in the horizontal row has twice as many sporozoites (i.e., 8, 16, 32, 64, etc.). The horizontal rows represent natural systematic groups and form homologous rows. The number of sporocysts in each such group is constant, but the number of sporozoites changes. In each such horizontal group there are homologous rows with the same number of sporozoites in the sporocysts. For example, in the two-sporocyst oocyst row we find the genus *Cyclospora* with two sporozoites in each sporocyst, matched by the genus *Eimeria* in the four-sporocyst row, which has two sporozoites in each sporocyst. We find that *Cryptosporidium* with four sporozoites is in the non-sporocyst row, matched by *Mantonella* in the one-sporocyst row, *Isospora* in the two-sporocyst row, and in the four-sporocyst row, *Wenyonella*, which has four sporozoites in each sporocyst. It is apparent from this table that not all the places have been filled, so far. During the 30-year period following Hoare's compilation of the table, the following genera were found: *Mantonella* from *Peripatus* and the crab, with four sporozoites per sporocyst, *Yakimovella* from the hedgehog, with eight sporocysts and a multitude of sporozoites, and *Hoarella* and *Pythonella,* each of which forms 16 sporocysts. The former has two sporozoites in each sporocyst, and the latter has four in each sporocyst (Fig. 31). Other vacant spaces in the table may eventually be filled.

## Reduction Division

For a general presentation of the developmental cycle of coccidia, it is necessary to determine the place and nature of reduction division. In the description of the general characteristics of the sporocysts it was noted that they are characterized by zygote reduction. Therefore, all stages in the developmental cycle, except for the zygote, have a haploid chromo-

some number. This conclusion was made on the basis of detailed cytological analyses of *Gregarina* and several representatives of the Coccidiomorpha. With reference to the latter, zygotic reduction was determined
for *Aggregata* (Dobell and Jameson, 1915; Dobell, 1925; Belar, 1926a,
1926b) and *Barrouxia schneideri* (Wedekind, 1927) which are representatives of the order Coccidiida and for *Karyolysus* (Reichenow, 1921a
and b), *Klossia helicina* (Naville, 1927), *Adelea ovata* (Greiner, 1921),
*Adelina cryptocerci* (Yarwood, 1937) and *Adelina deronis* (Hauschka,
1943) which are representatives of the order Adeleida. In recent years,
Grell (1953) and Bano (1959) observed zygote reduction in *Eucoccidium
dinophili* (order Protococcidiida) and the malarial parasites, respectively.

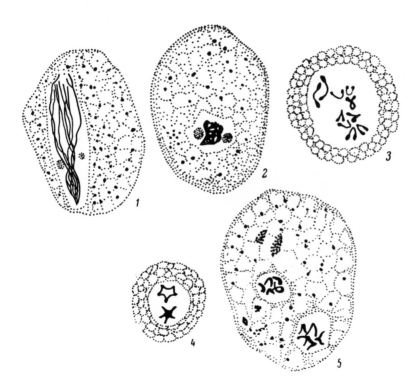

Fig. 32. Reduction division in *Barrouxia schneideri* (After Wedekind, 1927). *1*, Formation of the chromosome filaments during the prophase of metagamic division
(synkaryon spindle); *2*, formation of the solid tangle; *3*, diakinesis (10 paired
chromosomes); *4*, anaphase of the first division; *5*, prophase of the second division (five chromosomes in each nucleus).

In representatives of the family Eimeriidae reduction division has been observed only in *Barrouxia*. In *Eimeria, Cyclospora,* and others, this process has not been studied in detail. This can be explained by the fact that all coccidia are, in general, extremely small for cytologic studies and particularly so for study of the nuclei. Also, the walls which form around the zygote hinder staining of the nuclei. Thus, it is no accident that investigators have turned to the largest representatives of the coccidia, i.e., the various species of *Aggregata,* in which all processes of nuclear division are distinctly pronounced and visible to the investigator. *Barrouxia schneideri* from the intestine of *Lithobius* has also proved convenient for cytologic studies.

The individual stages of reduction division which have been observed in various coccidia are very similar to one another, so *Barrouxia* will be used to give a general picture of this process (Fig. 32). The synkaryon is situated in the center of the zygote. A large number of small granules of chromatin are scattered about in the nucleus around the large karyosome. The synkaryon stretches in preparation for the first metagamic division, and the chromatin takes the form of delicate filaments consisting of small granules. These filaments are oriented towards the poles of the stretched synkaryon. At this time, dissolution of the nuclear wall is observed. The chromatin filaments are highly stretched, and at this point it is possible to observe the long, thread-like chromosomes. This stage corresponds to the leptonema and is still sometimes called the spindle-fertilization stage, or even as Wedekind has suggested, the synkaryon spindle. It is impossible to distinguish the material obtained from the male or female nucleus of the stretched thread-like chromosomes. It is also very difficult to count the number of despiralized chromosomes at this stage. Apparently, an analogous synkaryon spindle was observed in the zygotes of *Eimeria stiedae* and *E. falciformis* by Reich (1913), who, however, did not attribute to it the appropriate significance. Ultimately, the spindle of the synkaryon begins to shorten, and each of the delicate chromosomes begins to thicken and form a tangle composed of the interwoven chromosomes. This is the pachyneme stage in which, after conjugation, the paired chromosomes form short, thick threads. In some cases the "bouquet" stage has been observed somewhat earlier.

During the pachyneme stage, the duality of each filament is clearly visible. Later it is possible to observe the final stage of prophase (diakinesis). At this time the chromosomes are very short and it is possible to see that each is split lengthwise. In *Barrouxia* five pairs of such chromosomes are visible at this point, but in *Aggregata* there are six pairs. The chromosomes are prominent at this point. During metaphase the shortened chromosomes become arranged at the equator of the spindle, and during anaphase their transmutation into the reduced number occurs. In the second division, which is typical mitosis, it is just as easily seen that *Barrouxia* forms a total of five chromosomes on the equatorial plate of metaphase; *Aggregata* forms six (Fig. 33). All subsequent metagamic divisions are equational and only the first division is by meiosis (Fig. 34). The nuclei of the sporozoites which are formed during sporogony are already haploid (Wedekind, 1927).

In some representatives of Adeleida and Coccidiida the number of chromosomes has been determined during division. They generally do not have many chromosomes, and the haploid number is from 2 to 8. For example, *Adelea ovata* has 4 or 5 chromosomes (Greiner, 1921); *Adelina cryptocerci* has 8 (Yarwood, 1937); *Klossia helicina,* 4 (Naville, 1927); *Klossia virrinae,* 4 (Grassé, 1953); *Klossia loosi,* 4 (Nabih, 1938); *Karyolysus bicapsulatus* and *K. biretorus* have 4 or 5 each (Reichenow, 1921a); *Barrouxia schneideri,* 5 (Wedekind, 1927); *Aggregata eberthi* and *A. octopiana* have 6 each (Dobell, 1925); *Eimeria maxima,* 5 (Scholtyseck, 1963a); *Isospora felis, E. bovis,* and *E. tenella* have 2 each (Walton, 1959); *E. debliecki* has 4 (Vetterling, 1966); and the various species of *Plasmodium* have 2 to 4 (Bano, 1959).

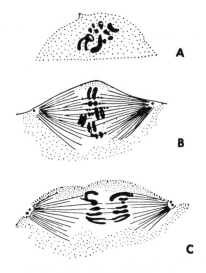

FIG. 33. Reduction division in *Aggregata eberthi* (After Belar, 1926b). *A,* Diakinesis; *B,* metaphase; *C,* anaphase.

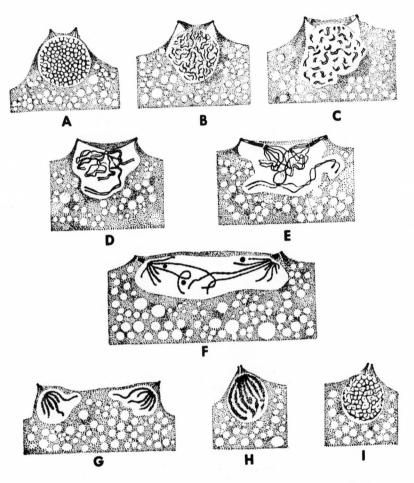

FIG. 34. Second metagamic division in *Aggregata eberthi* (After Dobell, 1925). *A*, Interphase with two centrioles; *B* and *C*, prophase; *D*, anaphase; *E*, late anaphase; *F*, transition to telophase; *G*, telophase; *H*, restoration of the nucleus and division of the centriole; *I*, interphase.

## Sex Determination

The presence of zygote reduction in sporozoans makes it possible to hypothesize as to the genotypic determinations of the sex of these protozoa. It may be assumed that male and female sex factors, which split off during reduction division, are concentrated in the nucleus so that the haploid products formed as a result of subsequent metagamic divisions must have half male and half female traits. The first two nuclei formed after the first metagamic division (reductional) should have different sexual potentials, i.e., both male and female. The sporoblasts each contain one such nucleus so that subsequent offspring should be of one sex. For example, in *Isospora,* division of the synkaryon into two nuclei leads to the formation of two sporoblasts. Each should have either a male or female potential. Because subsequent divisions are equational, the four sporozoites which form in each sporoblast will also be of one sex. If this is true, all offspring of one *Isospora* sporocyst should form either male or female gamonts while developing in the host organism. Likewise, in *Eimeria* each sporocyst with two sporozoites should also be of one sex. In such oocysts two sporocysts are male and two are female.

Inoculation of an animal with one *Eimeria* oocyst leads to an infection. Several investigators (Tyzzer, 1929; Jones, 1932; Roudabush, 1937; Kheysin, 1939, 1940, 1947b; Becker et al., 1955, 1956 and others) have found this by experimentation with coccidia of chickens, rats, rabbits, and other animals. Thus, the oocyst is a bisexual form, and whenever one oocyst enters the host it forms both macro- and microgametes. In any case, the cycle always ends with the formation of a new generation of oocysts.

There is little evidence of the genotypic determination of sex in coccidia. As an example, I might introduce Reyer's (1937) work with *Barrouxia.* Inoculation with one oocyst (containing a multitude of sporozoites) in the centipede *Lithobius forficatus* was successful. It was more difficult to infect the centipede with one sporocyst. Each *Barrouxia* sporocyst contains one sporozoite. Hence, the numerous sporocysts should be half male and half female. In some cases in which one sporocyst was used for infection only the macrogamete or microgametocytes actually developed in the intestine of *Lithobius,* and the full cycle of development did not take place when one sporocyst was used. These observations show that the sporozoites are monosexual, i.e., genotypic determination of sex takes place. If sex determination were phenotypic, then one sporocyst of *Barrouxia* would have formed both macro- and microgametocytes; however, this did not occur.

If the sporozoites are sexually differentiated, it could be expected that there would be sexual dimorphism among the asexual generations formed from sporozoites with different sexual potentials. In some species of coccidia, differences in the structure of schizonts and merozoites, of a type which might be ascribed to sexual differentiation, have been observed. Schaudinn (1902) described male and female schizonts in *Cyclospora caryolytica.* The large merozoites obtained from the large schizonts form the macrogametes, and small ones, from the smaller schizonts, give rise to the microgametocytes. However, Tanabe (1938) observed only one type of schizont in this species. He hypothesized that Schaudinn observed a mixed infection of species of *Eimeria* and *Caryospora*.

Because each species has several asexual generations, coccidia of rabbits, chickens, and other animals have a large variety of schizonts and merozoites. Whether such variation depends only on generations of schizonts and merozoites or whether there is also a sexual differentiation is difficult to prove because this question must be experimentally determined. Ikeda (1914) and Reichenow (1932) noted sexual dimorphism in the merozoites of *E. tenella,* but Scholtyseck (1953) could not confirm their data. In *E. magna,* some third-generation merozoites had a karyosome 1 micron in diameter and others measured 0.5 microns. It is possible that this is an expression of sexually differentiated merozoites (Kheysin, 1947b).

The treatment of the differences found in schizonts and merozoites may be rather arbitrary. Lainson (1965) found three types of schizonts and merozoites in *Cyclospora niniae* from the snake, *Ninia sebae sebae.* He hypothesized that one type of merozoite formed a new generation of schizonts, while the other two types formed macro- and microgametocytes, respectively. This has not been confirmed by direct observation.

Two types of merozoites, differing from each other in the amount of stored nutritive matter and nuclear structure, have been found in *Barrouxia schneideri.* Some merozoites are narrow and long and contain single small granules of carbohydrates in the cytoplasm. No nucleoli were found in the nuclei. Canning (1962) thought these formed microgametocytes in which there are also few granules of carbohydrates. Other merozoites, which were short and broad, had numerous granules of carbohydrates, and a large nucleolus was visible in the nucleus. Macrogametes, with large supplies of carbohydrate, are formed from these.

Much indirect evidence is in favor of genotypic determination. Development of macro- and microgamonts very often occurs in the same cell. In phenotypic determination of sex, the merozoites which enter one cell are usually under the influence of similar environmental conditions which would influence development into macro- or microgamonts; however, this does not happen.

It is somewhat surprising that in every species the number of macrogametes exceeds the number of microgametocytes. Counts of *E. intestinalis* in sections of various parts of the infected villi showed that the number of macrogametes always exceeded that of the microgametes by

10 or 15 times. It is highly probable that in the presence of sexual differ-entiation of asexual generations, the schizonts with differing sexual tendencies form different numbers of merozoites. Therefore, the differ-ence in numbers of macrogàmetes and microgametocytes in no way contradicts the possibility of genotypic determination of sex, but actually confirms it to a large degree.

# CHAPTER III

# Localization of the Various
# Stages of the Coccidian
# Life Cycle Within the Host

In the characterization of the suborder Eimeriidae, it was noted that the majority of the representatives are located within cells. The schizonts, gamonts, zygotes, and early stages of the oocyst usually develop in the host's cells. The merozoites develop only within the cells, after which they spend some time in extracellular life. The newly formed oocysts abandon the cells and enter the external environment, either without having begun or after completion of sporulation.

Representatives of the genera *Eimeria, Isospora, Wenyonella, Dorisiella, Cyclospora,* and others most often develop in the epithelial cells of various organs of vertebrate and invertebrate animals. However, some stages of coccidia develop in cells of mesodermal origin. The overwhelming majority of coccidian species are in the cytoplasm of cells and only a few are found in the nucleus. Examples of intranuclear parasites are *Cyclospora caryolytica* in the intestine of the mole, *Isospora mesnili* from the chameleon, *Eimeria ranarum* from the intestine of the frog, *E. salamandrae* and *E. grobbeni* from the intestine of the salamander, and *E. alabamensis* from the intestine of cattle.

Some species are not in cells, but in some kind of cavity not connected with the cells or on their surface, such as in the striated border of the intestinal mucosa. Apparently, the location of coccidia in such close contact with the epithelial cells has some sort of explanation. According to Ugolev (1963), adjacent digestion occurs on the surface of the epithelial cells of the intestinal mucosa. A high concentration of enzymes and the various substances that have been digested and absorbed through the microvilli are present at the surface of the epithelial cells of the villi. Under such conditions the endogenous stages of coccidian development can easily make use of the various substances which are needed for protein synthesis. An extracellular or pericellular location is not a permanent site for all stages of development. Sometimes gamonts, schizonts, or zygotes submerge into epithelial cells (Fig. 35). Very few coccidia are found under such conditions. This probably can be explained by the fact that the area on the surface of the cells is very small and there is always the possibility of being eliminated from the intestine.

Examples of pericellular coccidia (Fig. 36) are *Cryptosporidium muris* and *C. parvum,* which are parasitic in the intestine and stomach of mice, or some representatives of the genus *Eimeria,* e.g., *E. mitraria* from the intestine of the turtle, *E. pigra* from *Scardinius erythrophthalmus,* and *E. anguillae* from *Anguilla anguilla.* A common feature of these species is that the schizonts, gamonts, and oocysts are very small. Also, all stages of development are to some degree submerged in the cytoplasm of the epithelial cells. In *E. anguillae* the macrogametes protrude only slightly into the epithelium. Fertilization occurs outside

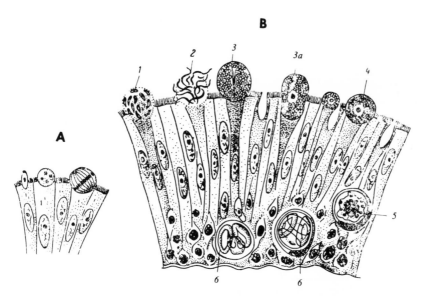

Fig. 35. Location of *E. anguillae* in the intestine of *Anguilla* (After Léger and Hollande, 1912). *A,* Schizogony; *B,* gametogony and sporogony; *1,* microgametocyte; *2,* microgametes; *3,* zygote; *3a,* submergence of the zygote into the epithelium; *4,* macrogamete; *5,* start of oocyst sporulation; *6,* sporulated oocysts.

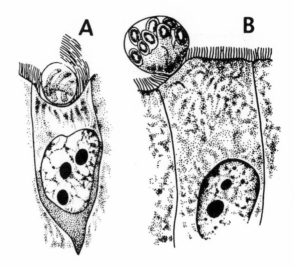

FIG. 36. Location of *Cryptosporidium parvum* in the intestine of a mouse (After Tyzzer, 1912). *A*, Mature oocyst with four sporozoites; *B*, schizogony.

the cell on the surface of the epithelium. The zygote is submerged in the cytoplasm of the epithelial cell where it develops into an oocyst. The gamonts of *E. brevoortiana,* a parasite of the American herring, begin their development outside the cell, then submerge into the cells of the pyloric appendages of the intestine where fertilization and oocyst formation take place.

*Eleutheroschizon duboscqui,* a coccidium of uncertain systematic position, lives in the intestines of polychaetes ( *Theostoma oerstedi* and *Scoloplos armiger*); the endogenous stages are on the surface of the epithelial cells of the intestine. They enter the lumen of the colon and are carried from the host into the water with the feces. The microgametes develop in the water, and fertilization and oocyst development occur there (Chatton and Villeneuve, 1936).

Most representatives of the Eimeriidae, regardless of whether they are in or outside of cells, parasitize the digestive tracts of vertebrate or invertebrate animals. Therefore, there is an excellent basis for calling the Eimeriidae intestinal parasites.

The endogenous stages of *E. percae* (Yakimov, 1929) develop in the epithelium of the intestinal mucosa of the perch. The endogenous stages of *Cryptosporidium muris* are on the surface of the epithelium of the stomach glands of the mouse (Tyzzer, 1910).

Some species of coccidia are in the liver. Among the coccidia of mammals, *E. stiedae* develops in the bile ducts of the liver of rabbits. A few species of coccidia of reptiles (lizards) develop in the epithelium of the gall bladder and the bile ducts, e.g., *E. legeri, E. agamae, E. scinci, E. flaviviridis, E. cystisfelleae, E. noctisauris, E. rochalimai,* and *E. acanthodactyli* (Bovee and Telford, 1965a and b). Among the fish coccidia, a group of species has been described whose oocysts are found in the liver. However, the student should approach with caution any data on location of coccidia based on the oocysts being found in any one organ. Reliable data on location can be obtained only by observing the endogenous stages of development in the proper organ. So in those cases in which there are no such data, it is more proper to speak not about localization in the given organ, but to say only that the oocysts have been found there. It is highly probable that such findings will in some cases reflect the actual localization of endogenous stages there.

In the livers of fishes, we have found oocysts of *E. cruciata* in *Caranx trachurus, E. clupearum* in the Pacific herring, *E. gasterostei* in the three-pronged stickleback, *E. auxidus* in *Auxis maru* and *Cololabis saira, E. pneumatophori* in the Japanese flounder, *E. metschnikovi* in the long-snouted Amur gudgeon, *E. amurensis* in *Pseudorasbora parva,* and *E. percae* in *Perca fluviatilis.* The endogenous stages of *Pfeifferinella impudica* are in the livers of mollusks.

A few coccidian species have been found in the kidneys. The endogenous stages of *E. truncata* are in the kidneys of geese. The oocysts of *E. metschnikovi* were found in the kidneys of the gudgeon *(Gobio gobio),* and *E. minuta* was observed in the kidneys of *Tinca vulgaris.* Shul'man and Zaika (1962) reported that oocysts of *E. strelkovi* were in the kidneys of the Amur chebak *(Pseudorasbora parva),* and Dogel' (1948) found oocysts of *E. sphaerica* in the kidneys of *Oristhocentros.* The oocysts of *E. amurensis* have been found not only in the liver, but also in the kidneys of *Pseudorasbora parva* (Dogel' and Akhmerov, 1946). *Isospora lieberkuehni* developed in the kidneys of frogs. During the spring the schizonts were in the glomerular epithelium of the kidneys of young frogs, and during the summer the gamonts developed in the epithelium of the renal canals. The oocysts also developed there and were excreted into the water with the urine (Nöller, 1923).

The endogenous stages of *E. gadi* are in the connective tissue of the wall of the swim bladder of codfish *(Gadus morrhua).* The oocysts of *E. macroresidualis* are found in the spleen of the Amur gudgeon (Shul'man and Zaika, 1962).

The oocysts of some species from fish have been observed in the testes, e.g., *E. sardinae* from *Clupea pilchardus* and *E. brevoortiana* from *Brevoortia tyrannus*. Apparently, the coccidia do not actually develop in this organ; the oocysts accumulate there and are eliminated from the host with the sperm. According to Hardcastle (1944), schizogony, gametogony, and fertilization of *E. brevoortiana* took place in the epithelium of the pyloric ceca of an American herring. The zygote that developed was motile and migrated from the intestine into the testis, where the oocysts were formed. Migration was from the intestinal wall into the body cavity. Hence, only an accumulation of oocysts actually occurred in the testes, but the entire developmental cycle took place in the digestive tract. *Caryotropha mesnili* develops in the body cavity of the polychaete, *Polymnia nebulosa*. The stages of development which are thought to be schizonts are in the spermatogonia swimming in the body cavity of the worm. The gamonts develop outside the cells.

Among the coccidia of mammals there are many references to locations outside the intestine (see, for example, Yakimov, 1931). According to Yakimov (1931) the oocysts of *E. stiedae* were found in the lymph nodes of a rabbit. He also found these same oocysts in the mesenteric nodes of the rabbit, in the bladder, and in the nasal passages. In the last instance it was supposed that coccidia even cause rhinitis in rabbits. However, a careful study revealed that the coccidia cannot be regarded as the cause of this illness, which is of viral origin (Smetana, 1933a, b, and c; Metelkin, 1936). Vagin (1930) reported oocysts of *E. stiedae* in the spleen.

Losanov (1963) reported that coccidia of chickens may localize in the liver, spleen, and brain. Lotze et al. (1964) observed schizonts of *E. arloingi, E. ninakohlyakimovae,* and *E. faurei* in the mesenteric lymph nodes of sheep 13 to 18 days after infection. The schizonts reached a diameter of 30 to 360 microns; however, no mature merozoites were found. The authors hypothesized that sporozoites might penetrate along the lymph ducts into the nearest nodes where schizogony began. Such a location cannot be considered normal.

Probably, the cases in which coccidia have been found in the lymph nodes, bladder, nasal cavity, brain, and spleen of chickens and rabbits, partially represent annoying mistakes by the investigator. In the first place, accidental contamination of the organ by oocysts from the intestine is possible. Secondly, in the study of histologic preparations of any organ, some of the host cells may be mistaken for coccidia, as undoubtedly happened in Losanov's research and probably in Lotze's as well. More careful experimentation is needed in order to resolve the question of the possibility of a change in location of coccidia in the host.

Recent work on cultivation of endogenous stages of coccidia forces us to give more attention to certain data on the finding of endogenous stages of coccidia in cells other than those of the intestine. The difficulty is that in tissue culture the asexual generations and gamonts of coccidia may develop in cells of different origins. Patton (1965) observed the development of asexual stages of *E. tenella* from chickens in a monolayer culture of fibroblasts obtained from 9-day-old chick embryos. When sporozoites were placed in such cultures they penetrated fibroblasts. They underwent schizogony for 4 to 6 days inside these cells. In chicks this process lasts no longer than 3 days. The increase in schizont development is probably the result of the unnatural conditions of the parasite's existence in the culture cells from connective tissue. What is surprising is not the retarded development, but the general ability of the schizonts to develop in cells which under natural circumstances never harbor the parasite. Even more surprising is the fact that schizont development may also occur in a monolayer culture of fibroblasts and kidney epithelium from cattle. Coccidia, in general, have a narrow host specificity. For example, no one has ever succeeded in using chicken coccidia to infect either mammals or other species of birds (Becker, 1934). In tissue culture, the cells isolated from the whole organism apparently lose their specific properties, and as a result, the parasite has a chance to develop in them without encountering any protective mechanisms which act in the whole organism.

An analogous phenomenon was also observed in the in vitro cultivation of the endogenous stages of *E. acervulina* from chickens (Strout et al., 1965). In tissue cultures, the sporozoites of this species penetrated into different cells. They were observed for several hours, after introduction into the culture in the fibroblasts from a chick and mouse, in kidney cells from chickens, in cells from a human amnion, and in He La cells. A light space in the cytoplasm of all these cells was noted around the sporozoites. Strout and his colleagues observed the process of schizogony in these cells which resulted in the formation of merozoites.

Long (1965) obtained complete development of the endogenous stages of *E. tenella* in chorioallantoic cells from 10-day-old chick embryos. The sporozoites were introduced into the allantoic cavity. During the next 7 days, schizont development was observed. By the 7th day adult schizonts were present; these were similar to the second-generation schizonts of *E. tenella*. During the next 4 days, gamonts and oocysts were seen in the chorioallantois; they were found not only in the cells, but also free in the allantoic fluid. The oocysts ultimately sporulated normally and caused an infection in week-old chicks. Thus, the endogenous development of *E.*

*tenella* may take place not only in the cells of the ceca but also in vitro in cells which have a completely different function and origin.

Usually each species of coccidium localizes in a single organ (e.g., *E. stiedae* only in the liver, *E. truncata* only in the kidneys, and *E. media, E. acervulina,* and other species only in the small intestines). But there are data reporting oocysts of one species in more than one organ. On this basis, conclusions have been made as to the localization of endogenous stages of a single species in various organs. An example might be some species of fish coccidia. The oocysts of *E. cheissini* were found in the mesentery (which is highly doubtful), gall bladder, swim bladder, and intestine of the long-snouted Amur gudgeon (Shul'man and Zaika, 1962). The oocysts of another species *(E. metschnikovi)* are found in the

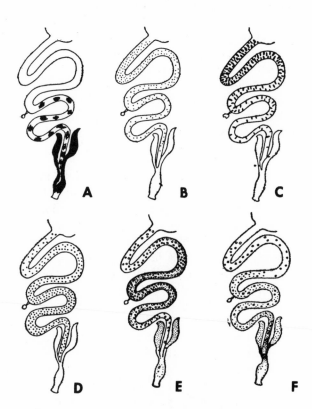

FIG. 37. Diagram of the location of six species of chicken coccidia. *A, E. tenella; B, E. mitis; C, E. acervulina; D, E. maxima; E, E. necatrix; F, E. brunetti;* (*A* to *E,* After Tyzzer, 1929; Tyzzer et al., 1932; *F,* After Boles and Becker, 1954).

same host, but in the liver, kidneys, spleen, and intestine. Finally, the oocysts of *E. siliculiformis* from the same host are found in the wall of the swim bladder, in the intestine, and the kidneys. Of course, such variety in localization of the same species is very surprising, if, of course, no errors in methodology have been allowed to enter into the research. The probability has not been eliminated that accidental contamination of the organ might have taken place and confused the investigator as to the actual location of infection.

Most coccidia are in a definite organ or even a definite part of an organ. Among the intestinal coccidia, there is a group of species whose development occurs in one part of the intestine in one stage and in another area of the intestine during the next stage. For example, in *E. coecicola*, a coccidium of rabbits, the asexual generations develop in the small intestine, but the gamonts develop in the appendix and cecum (Fig. 39). In *E. necatrix* the schizonts of the first two generations develop in the middle part of the small intestine, but the third-generation schizonts and the gamonts are in the ceca (Fig. 37). Among the coccidia of turkeys, the first-generation schizonts of *E. meleagridis* develops in the small intestine, but the second generation and the gamonts develop in the ceca (Fig. 38). In cattle, first-generation schizonts of *E. bovis* develop in the small intestine, but second-generation schizonts and gamonts develop in the large intestine.

It is sometimes observed that schizonts and gamonts may simultaneously localize in various parts of the intestine. For example, in the rabbit [schizonts and gamonts of] *E. intestinalis* are located in the lower part of the small intestine, but the gamonts also develop in the cecum

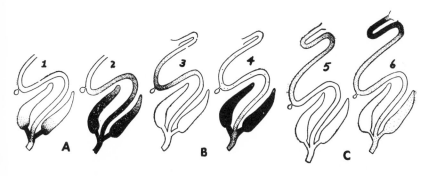

Fig. 38. Diagram of the location of turkey coccidia (After Clarkson, 1958, 1959a and b). *A, E. adenoides; B, E. meleagridis; C, E. meleagrimitis; 1,* location of asexual generations; *2,* gamonts.

and appendix. Thus, oocysts form in both the small and large intestines. The endogenous stages of *E. brunetti* in chickens were found in the lower part of the small intestine, the ceca, the colon, and the cloaca (Boles and Becker, 1954). In the same host, schizogony of *Eimeria mivati* took place in the upper third of the small intestine, but the gamonts were in the lower part of the small intestine, the ceca, and the colon (Edgar and Seibold, 1964). *Wenyonella gallinae* is located in the lower part of the small intestine and in the colon of chickens. All stages of development of *E. adenoeides* occur in the lower part of the small intestine and in the ceca and colon of turkeys.

Usually coccidia are located in a single organ, and the greatest concentration of the endogenous stages of the parasites is found in a single part of that organ (Fig. 39). *Eimeria media, E. irresidua,* and *E. perforans* develop only in the small intestine of the rabbit, but *E. piriformis* is found in various parts of the large intestine. Each species has a most characteristic and limited part of a given organ where the greatest number of endogenous stages develop. For example, the greatest numbers of *E. media* are found in the duodenum and the upper part of the

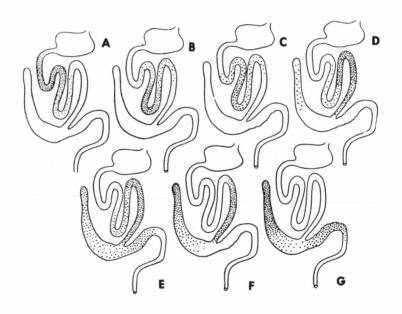

Fig. 39. Diagram of the location of intestinal coccidia of rabbits (After Kheysin, 1957b). *A, E. media; B, E. irresidua; C, E. perforans; D, E. magna; E, E. intestinalis; F, E. coecicola; G, E. piriformis.*

small intestine of rabbits, but sometimes the endogenous stages of this species are found as far as the middle of the small intestine. *Eimeria perforans* is concentrated in the middle part of the small intestine of rabbits, but it is also found from the duodenum to the lower part of the intestine. *Eimeria irresidua* concentrates in the upper third of the small intestine, and *E. magna* is found in the lower part of the small intestine 10 to 20 cm from the cecum, although sometimes, especially in an intense infection, the endogenous stages spread over about half the length of the small intestine (Kheysin, 1947b). In rabbit coccidia (as in coccidia of chickens), one can observe that the more intense the infection, the wider the area in which the endogenous stages of the species is likely to settle.

*Eimeria maxima, E. acervulina, E. mitis, E. praecox,* and *E. hagani* (Fig. 37) develop in the small intestine. *Eimeria maxima* has the largest area of development. The endogenous stages of this species, however, are concentrated in the middle and upper part of the small intestine. *Eimeria acervulina, E. mitis, E. praecox,* and *E. hagani* are located in the anterior half of the small intestine; the last species develops primarily in the duodenum, and the other three in various parts of the jejunum. *Eimeria praecox* is the nearest to the duodenum. *Eimeria tenella* is found in the ceca, but in rare cases the endogenous stages have been observed in the lower part of the small intestine and colon.

*Eimeria kotlani* is found in various parts of the large intestine of geese. The endogenous stages of this species develop primarily in the colon, the cloaca, and in the terminal (thin) parts of the ceca. *Eimeria meleagridis* develops only in the small intestine of turkeys. It concentrates in the duodenum and the upper part of the small intestine. In rare cases the gamonts are observed in the ceca and colon (Fig. 38).

The endogenous stages of *E. auburnensis, E. ellipsoidalis, E. alabamensis,* and *E. bukidnonensis* are found in various parts of the small intestine of cattle. They are concentrated in the lower part of the small intestine; however, large numbers of *E. auburnensis* are also found 2 to 6 meters below the ileocecal valve. It is usually observed that as the parasites develop they spread more widely over a given organ. For example, the first- and second-generation schizonts of *E. intestinalis* occupy about 15% of the length of the small intestine of month-old rabbits. The third-generation schizonts and the gamonts occupy about 40% of the length of the small intestine and are also found in the large intestine (Kheysin, 1948). First-generation schizonts of *E. debliecki* occupied 15% of the entire length of the small intestine of pigs. They were found 30 to 180 cm from the stomach (when the total intestinal length is 914 to

1065 cm). Second-generation schizonts were distributed even farther down the intestine and reached 400 cm from the stomach. Forty-three per cent of the small intestine was infected by second-generation schizonts. The gamonts occupied 70% of the small intestine and were found up to 650 cm from the stomach (Vetterling, 1966).

*Eimeria separata* in rats and *E. caviae* in guinea pigs develop only in the large intestine and cecum. Two species, *E. miyairii* and *E. nieschulzi* are located in the small intestine of the rat. Species of *Isospora* of cats and dogs are found only in the small intestine.

The endogenous stages of development of a given species localize in a definite organ or in a definite part of an organ. Some species have a wider distribution, others a rather narrow one. However, those species which are adapted to development in a definite organ do not develop in other organs. Numerous experimental infections of rabbits with the liver coccidium, *E. stiedae,* have shown that the endogenous stages of this species will not develop in the intestine, nor will the intestinal species develop in the rabbit's liver. *Eimeria piriformis,* which always localizes in the large intestine of the rabbit, has never been found in the small intestine (Kheysin, 1963). Apparently, such a strict adaptation of endogenous stages to life in a definite organ or organ part has been determined by the character of the interrelationship between the intracellular parasites and their surrounding environment. The morphophysiological properties of the tissues and cells of a given organ or its parts apparently have a great effect on the development of coccidia. If these properties are suitable for the existence of the species, then development will occur in the proper cells and tissues.

Herrick (1936) found experimentally that coccidia can develop in cells of a given organ even if the normal function of the organ was changed. The ceca were removed from the normal place in a chick and transplanted to an upper region of the small intestine where they functioned as part of the small intestine. When the chick was inoculated with *E. tenella* oocysts, the oocysts developed in the transplanted part of the ceca. Herrick concluded that localization of the parasite in a definite organ was governed by an affinity between the endogenous stages of the developing coccidia and that tissue in which the species develops. Ikeda (1957) repeated these experiments. He transferred pieces of the ceca to the upper region and distal loop of the duodenum, and the middle region of the small intestine. In an *E. tenella* infection, the endogenous stages developed only in the cecal portion attached to the loop of the duodenum and middle part of the small intestine. Thus, he confirmed Herrick's conclusions concerning the tissue and cell specificity of coccidia.

Intestinal pH and other factors have a comparatively small effect on localization of coccidia in the gut. The locations of *E. magna, E. media,* and *E. irresidua* do not change if the rabbits are fed diets which alter the intestinal pH. On a milk diet the pH was 5.9 to 6.2, but on a protein and vegetable diet it was 6.4 to 6.8. The location of *E. magna* and other species remained unchanged under all of these conditions.

The intestinal coccidia are characterized not only by localization in a definite part of the intestine, but also in definite cells of the intestinal mucosa. Coccidia are usually found in epithelial cells, and in many species the entire cycle takes place in the epithelium. This type of development occurs in *E. perforans, E. intestinalis,* and *E. piriformis* from the rabbit (Fig. 40). Most species of chicken coccidia also develop in the epithelium, e.g., *E. necatrix, E. acervulina, E. mitis, E. praecox, E. brunetti, E. hagani, E. mivati* (Figs. 50 to 52, 54, and 55). *Eimeria adenoeides, E. gallopavonis, E. meleagridis,* and *E. meleagrimitis* (Fig. 56) all develop in the epithelium of the intestinal mucosa of turkeys. The same localization occurs in *E. separata* and *E. nieschulzi* of rats and *E. caviae* of guinea pigs. *Eimeria zurnii, E. auburnensis, E. ellipsoidalis, E. alabamensis,* and *E. bukidnonensis* from cattle also localize in the epithelium.

However, in some cases asexual generations and gamonts develop in the epithelium, but the oocysts submerge in the underlying connective tissue where development is completed. Such development is observed, for example, in *E. coecicola* in the appendix of the rabbit. The tendency for development of part of the endogenous cycle in the connective tissue is pronounced in various coccidia (Kheysin, 1957a and b). In *E. media* and *E. irresidua* from the rabbit, only asexual reproduction occurs in the epithelium, but gametogony, fertilization, and oocyst formation take place in the subepithelium. A similar development of *Isospora felis* and *I. rivolta* occurs in the intestines of dogs. According to Marinček (1965), the early stages of schizogony and the young macrogametes of *E. subepithelialis* are in the epithelium of the carp's intestine. These stages then submerge into the connective tissue, where development is completed. Fertilization occurs in the subepithelium, as does the formation of oocysts, which sporulate before they leave the host. The microgametocytes are found only in the epithelium. In *E. bovis* the first-generation schizonts develop in the subepithelium, but those of the second generation and the gamonts are in the epithelium.

In some species, development of schizonts and gamonts begins in the epithelium, after which the schizonts and gamonts submerge into the connective tissue and complete their development with the formation of

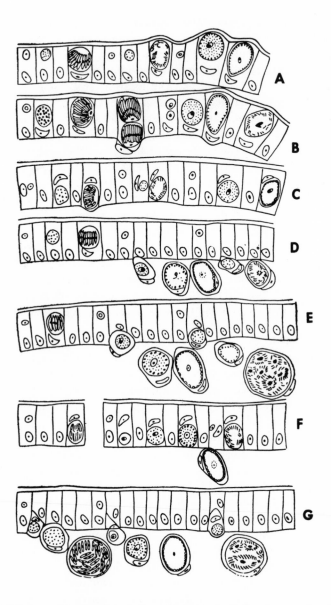

Fig. 40. Diagram of the location of the endogenous stages of various species of rabbit coccidia in the intestinal wall (After Kheysin, 1957b). *A, E. piriformis; B, E. intestinalis; C, E. perforans; D, E. media; E, E. irresidua; F, E. coecicola; G, E. magna.*

oocysts. Such development is observed in *E. magna* from the rabbit and *E. tenella* from chickens.

Tyzzer (1929) hypothesized that when the epithelial cells of the cecal mucosa of chickens were infected with coccidia, the cells actively migrated into the underlying connective tissue. This opinion was confirmed by Kheysin (1940, 1947b) during a study of the endogenous stages of *E. magna* and other coccidia of the rabbit. The infected epithelial cells are pushed out of the epithelial layer by the new cells growing in the crypts and submerge into the tunica propria. Different stages of this process can be observed in the intestine of rabbits infected with *E. magna*. The parasite remains in the epithelial cell, which has, however, changed its normal location. It is sometimes possible to see only partial movement into the subepithelium. Such cells retain a connection with the cells in the epithelial layer. In other cases, they separate completely from the epithelium and lie among the cells of the connective tissue.

The epithelial cells containing the endogenous stages of some coccidia are also found in connective tissue cells. According to Hammond, Clark, and Miner (1961), in cattle the gamonts of *E. auburnensis* localized under the epithelium of the villi in cells of mesodermal origin. First-generation schizonts of *E. bovis* develop in the endothelial cells of the lacteals of the villi of the small intestines of cattle.

Before the sporozoites of *E. necatrix* and *E. tenella* begin development in epithelial cells, they penetrate into macrophages, not only passively, but apparently actively as well (van Doorninck and Becker, 1957; Challey and Burns, 1959).

The endogenous stages may localize in various parts of the epithelium of the intestinal mucosa. In some cases they are found in the distal part of the villi, but in others they are located along the entire villus at the base of the crypt or in the epithelium of the crypt of both the small and large intestine. They may be present on the bottom or in the apical region of the crypts. For example, the schizonts and gamonts of *E. media* are in the distal region of the villi of the intestinal mucosa of rabbits (Fig. 41). The endogenous stages of *E. irresidua* and *E. intestinalis* develop along the entire villus and in the crypts, and *E. perforans* localizes near the bottom of the crypts of the large intestine. This is also observed among the coccidia of chickens. The second-generation schizonts of *E. brunetti* are in the distal part and along the entire villus, the gamonts of *E. maxima* develop in the central part of the villus, and the schizonts of *E. necatrix* are situated at the bottom of the crypts.

First-generation schizonts of *E. adenoeides* are found near the distal

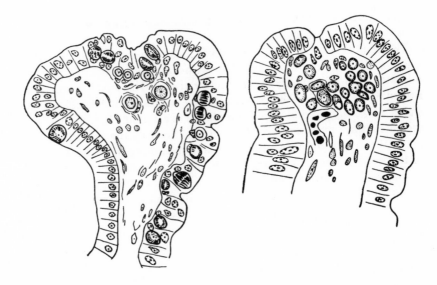

FIG. 41. Location of the schizonts and gamonts of *E. media* in the villus of the upper region of the rabbit intestine (After Kheysin, 1947b).

end of the villi, but second-generation schizonts and gamonts are along the entire villi and in the crypts. First-generation schizonts of *E. meleagridis* localize in the turkey's intestine at the base of the villi, but not in the crypts; however, second-generation schizonts and gamonts develop in the depths and surface of the crypts.

In cattle first-generation schizonts of *E. bovis* are present in the villi, but the second-generation schizonts and gamonts develop in the epithelium of the crypts. Since the merozoites often penetrate into the epithelial cells near the place where the merozoites are formed, large foci of schizonts (colonies) are observed in the epithelium. The formation of such colonies is usually observed in the development of the second-generation schizonts of *E. tenella, E. necatrix,* and other coccidia of chickens (Tyzzer et al., 1932), and in the coccidia of rabbits (Kheysin, 1947b). Some villi are heavily infected with schizonts, but adjacent ones may be completely free of parasites.

Various stages of intracellular development localize either above or below the nucleus. First- and second-generation schizonts of *E. tenella, E. maxima, E. acervulina,* gamonts of *E. acervulina, E. meleagridis, E. meleagrimitis,* schizonts of *E. gallopavonis, E. intestinalis, E. irresidua, E. media,* and schizonts and gamonts of *E. intestinalis, E. perforans, E.*

*magna* are located above the host cell nuclei. A subnuclear position is occupied by third-generation schizonts and gamonts of *E. tenella,* gamonts of *E. maxima,* schizonts and gamonts of *E. mitis* and *E. prae-cox,* first- and second-generation schizonts of *E. brunetti,* and second-generation schizonts of *E. gallopavonis.*

It is apparent from the above data that when several species of coccidia are present in one host, their endogenous stages are spatially isolated both along the length of the intestine and in the various divisions of the mucosa. As a result, when one host is simultaneously invaded by several species, each develops without suppressing the others. For example, the location of *E. magna* and *E. intestinalis* in the lower part of the small intestine overlaps, but the endogenous stages of the latter species develop only in the epithelium, while only the young schizonts and gamonts of *E. magna* begin development in the epithelium; infected cells with parasites rapidly submerge into the connective tissue, where final development of the endogenous stages takes place (Fig. 40). Thus, there is no competition for space between these two species. The same is true of several other pairs of species.

The gamonts of *E. coecicola* develop below the nuclei of epithelial cells of the appendix of rabbits; *E. intestinalis* develops above the nuclei and the gamonts of *E. magna* develop in the subepithelium. The gamonts of *E. intestinalis* and *E. piriformis* localize at varying depths in the crypts of the large intestine. The first species is more superficial than the second. The areas of development of *E. media, E. perforans,* and *E. irresidua* overlap somewhat in the jejunum. However, *E. media* occupies the distal part of the villi, while the two other species are closer to the base and even in the crypts. The gametes of *E. perforans* develop in the epithelium, but the other two species are in the tunica propria. Also, the schizonts and gamonts of *E. perforans* are below the nuclei, but the same stages of *E. media* and *E. irresidua* are located above the nuclei. These species may develop in the intestine of the rabbit at the same time without competing with each other (Kheysin, 1947b).

In some cases where two species are in the same place, an antagonism in their development does occur. For example, it was found that when a rabbit was simultaneously inoculated with oocysts of *E. coecicola* and *E. intestinalis* (which have an overlapping localization in the small intestine) the development of both species was suppressed. The prepatent period was retarded for 1 or 2 days, and the productivity of the parasite was found to be considerably lower than that occurring in separate development. It is possible that this was associated with the more intense asexual reproduction of *E. intestinalis.* The schizonts of this species hinder, to some degree, the development of the schizonts of *E. coecicola*.

An analogous spatial isolation is also observed in the coccidia of turkeys. This is partially expressed in the localization of parasites along various parts of the intestine. When places of development overlap, there is a separation of endogenous stages in the intestinal mucosa. For example, first-generation schizonts of *E. meleagrimitis* and *E. melagridis* may be found in the same part of the small intestine, but the first species is in the epithelium of the crypts; however, the second generation is along the entire villi but is not found in the crypts. The gamonts of *E. adenoeides* and *E. meleagridis* are found in the crypts of the ceca, but the gamonts of the former species can also develop in the lower part of the small intestine so that a spatial isolation of the stages occurs for these species. Schizonts of these species are separated along the length of the intestine; also, in *E. adenoeides* they are distal to the end of the villus, but in *E. meleagridis* they are at the base.

Some spatial isolation is also found in coccidia of chickens. Gamonts of *E. tenella* and *E. necatrix* are found in the cecum. However, the schizonts of the latter species develop in the middle part of the small intestine. *Eimeria acervulina* and *E. mitis* develop in the anterior part of the small intestine; however, schizonts and gamonts of the first species are above the nuclei, but those of the second species are below the nuclei of the epithelial cells.

Dogel' (1949) stated that the specific location of each species of coccidia in a single host reflects an autochthonic process of species formation (Kheysin, 1957b). This opinion is all the more probable because coccidia usually have a narrow host preference. It is possible that each host species which now harbors several species of coccidia had only one species at one time, and that its endogenous stages settled in various parts of a given organ, such as the digestive tract, and also in various parts of the mucosa. Later, several species were formed which differed only in their connection with a localized site of infection in a definite part of the intestinal tract. Territorial isolation of various stages of development within one host provided an important condition for species formation of the coccidia.

In every case, localization of endogenous stages of coccidia in different cells of the host leads to damage of the cells. Without examining this question in detail, we can only note that when the epithelial cells of the intestinal mucosa are infected by a parasite, they often become greatly enlarged and lose their usual form and association with neighboring cells. The host cells swell because of the parasite's growth and may even change their shape entirely. An example is the epithelial cells of the intestinal mucosa and the gall bladder of several lizards infected with *E. sceloporis* and *E. noctisauris* (Bovee and Telford, 1965a and b). The early stages of development of these species are under the external surface of the epithelial cells. As the schizonts or gamonts mature the cells infected by the parasite stretch and grow in the direction of the lumen of the intestine or the gall bladder. The epithelial cells infected by *E. nocti-sauris* remain united with the epithelium by a small "foot" (Fig. 42). The mature schizonts and gamonts are in epithelial cells and appear as if they are above the epithelium. As a result of the growth of the parasite, the apical part of the epithelial cells is greatly swollen and covers the coccidium with a thin layer (Fig. 42).

The infected cell always collapses after the parasite abandons it. The nuclei of the epithelial cells are severely deformed by schizonts of *E. intestinalis* and *E. magna*. They become sickle-shaped with the concave side usually lying toward the parasite. The nuclei of cells infected with *E. tenella*, *E. bovis*, and other *Eimeria* species have a similar appearance. Also, the nuclei are enriched with ribonucleic acid and lose deoxyribonucleic acid. Such nuclei give a weak Feulgen reaction (Kheysin, 1958a).

The cytoplasm of epithelial cells infected with gamonts of *E. perforans* shows an increase in the number and concentration of mitochondria between the nucleus of the cell and the parasite. Electron microscope

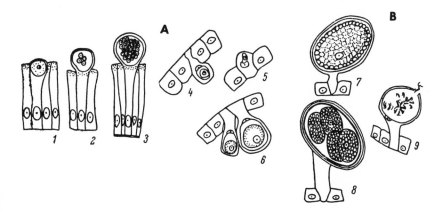

FIG. 42. Location of the endogenous stages of *E. sceloporis* (*A*) and *E. noctisauris* (*B*) in the intestine of lizards (After Bovee and Telford, 1965a and b). *1* to *3*, Schizogony; *4* and *5*, trophozoites in epithelial cells; *6*, macrogametes; *7*, zygote; *8*, oocyst with four sporoblasts; *9*, mature microgamete.

studies have shown that many mitochondria vacuolize and degenerate. Numerous fibrils appear in the cytoplasm of the host cell surrounding macrogametes of *E. perforans* (Scholtyseck, 1963b).

The endogenous stages of coccidia may disturb the metabolism of the cell. When schizonts and gamonts of *E. intestinalis* develop in the epithelium of the villi of the small intestine of rabbits, the activity level of alkaline phosphatase decreases sharply compared to that of cells of noninfected parts of the intestine. Alkaline phosphatase is localized in the striated border of the epithelial cells of the intestinal mucosa. The more intense the infection, the more clearly the disappearance of alkaline phosphatase from the striated border is expressed. The processes of digestion and absorption of material in the small intestine are disturbed in the same way (Beyer, 1960).

# CHAPTER IV

# The Continuity of a
# Coccidian Infection

Coccidiosis begins the moment that occysts enter the body and continues during the establishment of the endogenous stages until the host is completely free of the parasite. It is highly characteristic that a coccidian infection is limited in time. At least this is true of many species of the genus *Eimeria* and several species of the genus *Isospora*. The time limitation is evident if the host is infected only once by the oocysts of a given species of coccidium. The development of the infection does not continue indefinitely, but ends if no further infection occurs within a short time. Tyzzer (1929) studied this phenomenon in chickens, and observed that a coccidian infection was self-limiting in time.

What determines such a phenomenon? The endogenous development of coccidia, i.e., development within the host, consists of a sequential alternation of asexual reproduction followed by the sexual cycle which ends in fertilization and the formation of a zygote that becomes encysted to form an oocyst. The oocyst is eliminated into the external environment where exogenous development (sporogony) occurs. However, in some cases, sporogony may occur prior to the elimination of the oocysts into the external environment, but this does not alter the general scheme of the coccidian life cycle. In one way or another fertilization is always followed by the formation of an encysted zygote (oocyst) which sporulates, and sporozoites form within it. Sporulation of the oocysts within the host has been observed in *E. subepithelialis, E. carpelli, E. gadi,* and other coccidia of fish, and in *I. bigemina* from cats and dogs. Endogenous development is determined by the period of time from penetration of the sporozoite into the host organism until the formation of oocysts.

Exogenous development (sporogony) usually takes place outside the host.

The duration of endogenous development is determined by the total time needed to complete asexual reproduction, gametogony, fertilization, and oocyst formation. Of all the endogenous stages of development, only schizogony has the ability to repeat itself. Microgametogenesis results in the formation of microgametes, from which microgametocytes cannot be formed in any way. The same is true of macrogametogenesis. After fertilization, the newly formed macrogamete must become a zygote, then an oocyst, and it cannot repeat development to become a trophozoite. If gamonts are developed in the cycle, no return to asexual reproduction occurs. The same is true of sporogony. Here, development always occurs in one direction and ends in the formation of sporozoites. Merozoites are formed during schizogony, and these may give rise to schizonts of the next generation. Therefore, the continuity of a coccidiosis infection or the duration of the endogenous phase of the developmental cycle will be determined first of all by the number of asexual generations (as well as by how often they repeat), the length of the transition period from asexual to sexual reproduction, and by the duration of gametogenesis.

If schizogony were repeated indefinitely, a coccidian infection would be of unlimited duration. In a single infection by coccidia, the host would maintain the parasite, as indicated by the elimination of the oocysts, over an unlimited period. However, a study of many species of coccidia belonging to the genus *Eimeria* has failed to demonstrate any such development.

There is another possibility. Schizogony could repeat itself a limited number of times. In such a case the host would free itself of the parasite within a certain period of time. In such a development, elimination of the oocysts would not continue but would end rather quickly. Such a phenomenon has been observed among various coccidia; this indicates that the invasion is self-limiting in time.

A coccidiosis infection consists of two periods. The first lasts from the moment the host is infected to the formation and elimination of the first oocysts into the external environment. This is the prepatent period (pre—before, until; patent—exposed, manifest). The second period lasts from the appearance of the first oocysts until they completely disappear, i.e., until the end of the period during which occysts are passed by the host (if the parasite localizes in the intestine or liver). This is the patent period (Andrews, 1930).

## THE PREPATENT PERIOD

During the first period, the development of the coccidia in the host is hidden, and only the appearance of oocysts gives evidence of the existence of an infection. [A clinician would certainly not always agree with this statement. Ed.] The length of the prepatent period depends on how many asexual generations precede the appearance of the first gamonts and how long the asexual generations and gametogenesis last. For each species of coccidium these periods are more or less constant, and in any case they are not highly variable. Hence, the prepatent period is relatively constant for each species. Actually, in many species of coccidia the length of the prepatent period is a reliable specific characteristic. Table 2 gives data on the length of the prepatent period of many coccidia. Unfortunately, these data include no more than 5% of the known species of coccidia of the family Eimeriidae. This is because most species of coccidia are described only from the characteristics of the oocysts. Experimental infection of the host, which might help to determine the length of the prepatent period, does not in most cases accompany the description of a new species. As a result there are very few data on the prepatent period as compared with the number of known species. This is particularly true of coccidia of wild animals. The length of the prepatent period has been studied best in the coccidia of domestic mammals and birds. The developmental cycles of these coccidia have been studied experimentally by maintaining conditions which hinder spontaneous or accidental infection. If such conditions are not maintained, erroneous data on the length of the prepatent period will be obtained by the investigator.

In analysis of the data in Table 2, it is evident that in most cases the prepatent period is not long and in many species and hosts fluctuates somewhere between 4 and 11 days. None of the species examined had a prepatent period shorter than 4 days. Some species had longer prepatent periods. For example, the prepatent period of *E. stiedae,* which is found in the bile ducts of the rabbits, was 17 days (Kotlán and Pellérdy, 1949). Smetana (1933a) reported that the first oocysts of this species appeared within 3 or 4 weeks. My data confirm those of Kotlán and Pellérdy. Older data of several authors (Reich, 1913; Pérard, 1924; Waworuntu, 1924; Yakimov, 1931; Carvalho, 1943) record a 6- to 11-day prepatent period for *E. stiedae;* these are apparently the result of inaccurate experiments and the failure to observe the basic rules for protecting rabbits against spontaneous infection. A long prepatent period of 18 days or longer has

TABLE 2
Duration of the prepatient period of various coccidia
according to published data

| HOST | SPECIES OF PARASITE | PREPATENT PERIOD IN DAYS | PUBLISHED SOURCE |
|---|---|---|---|
| Cattle | Eimeria auburnensis | 18 (19–20) | Hammond et al., 1961 |
| | E. bovis | 18 | Hammond, 1964 |
| | E. zurnii | 19 (8–28) | Davis & Bowman, 1957 |
| | E. ellipsoidalis | 10 | Hammond et al., 1963 |
| | E. bukidnonensis | 25 | Davis & Bowman, 1964 |
| | E. alabamensis | 6–11 | Davis et al., 1955 |
| Sheep and goats | E. arloingi | 22 | Christensen, 1941 |
| | | 18, 19 | Lotze, 1953; Krylov, 1959b |
| | E. parva | 10–16 | Lotze, 1953 |
| | | 14–15 | Krylov, 1959b |
| | E. faurei | 15–16 | Ibid. |
| | E. ninakohlyakimovae | 10–13 | Balozet, 1932 |
| | | 15 | Shumard, 1957 |
| | | 14–15 | Krylov, 1959b |
| | E. intricata | 24 | Ibid. |
| | | 22–27 | Davis & Bowman, 1965 |
| | E. crandalis | 14 | Krylov, 1959b |
| | E. ahsata | 20–21 | Ibid. |
| Swine | E. debliecki | 7 | Deom & Mortelmans, 1954 |
| | | 5–6 | Pellérdy, 1949; Wiesen-hütter, 1962 |
| | | 6, 5 | Vetterling, 1966 |
| | E. polita | 8–9 | Pellérdy, 1949 |
| | E. scabra | 9–10 | Ibid. |
| | Isospora suis | 6–8 | Biester & Murray, 1934 |
| Cat | I. felis | 7–8 | Hitchcock, 1955; Tomimura, 1957; Nemeséri, 1960 |
| | | 5–6 | Andrews, 1926 |
| Dog | I. bigemina | 5–6 | Lee, 1934 |
| | I. canis | 11 | Nemeséri, 1960 |
| Rat | E. nieschulzi | 7–8 | Roudabush, 1937 |
| | E. separata | 5–6 | Ibid. |
| | E. miyairii | 6, 5 | Ibid. |
| Nutria | E. seideli | 14 | Pellérdy, 1960 |
| Rabbit | E. magna | 8–9 | Kheysin, 1940, 1947 |
| | E. media | 5–6 | Ibid. |
| | E. irresidua | 7–9 | Ibid. |
| | E. intestinalis | 10 | Kheysin, 1948 |
| | E. piriformis | 10 | Ibid. |
| | E. perforans | 6 | Kheysin, 1940, 1947b |
| | E. coecicola | 9–10 | Kheysin, 1947c |

*(Cont.)*

TABLE 2 (continued)

| Host | Species of Parasite | Prepatent Period in Days | Published Source |
|---|---|---|---|
| Rabbit—*continued* | *E. stiedae* | 17 | Kotlán & Pellérdy, 1949 |
| | *E. neoleporis* | 11–14 (12) | Carvalho, 1944 |
| Guinea pig | *E. caviae* | 11–12 | Henry, 1932; Lapage, 1940 |
| House mouse | *E. falciformis* | 5 | Cordero del Campillo, 1959 |
| Peromyscus leucopis | *E. leucopi* | 5–6 | von Zellen, 1959 |
| Dipodomys panamantinum mohavensis | *E. mohavensis* | 7 | Doran & Jahn, 1952 |
| Squirrel | *E. silvana* | 7 | Pellérdy, 1954a |
| | *E. sciurorum* | 7 | *Ibid.* |
| | *E. andrewsi* | 6 | *Ibid.* |
| | *E. mira* | 10–11 | *Ibid.* |
| Skunk | *E. furonis* | 6 | Hoare, 1927 |
| | *E. ictidea* | 7 | *Ibid.* |
| Chicken | *E. tenella* | 7 | Becker, 1952 |
| | | (138 hours) | Edgar, 1955 |
| | *E. necatrix* | 6–7 | Davies, 1956; Tyzzer et al., 1932 |
| | | (138 hours) | Edgar, 1955 |
| | *E. hagani* | 7 | Becker, 1952 |
| | | (138 hours) | Edgar, 1955 |
| | *E. mitis* | 5 | Becker, 1952 |
| | | (99–101 hours) | Edgar, 1955 |
| | *E. praecox* | 4 | Becker, 1952 |
| | | (84 hours) | Edgar, 1955 |
| | *E. maxima* | 6–7 | Becker, 1952; Scholtyseck, 1963a |
| | | (123 hours) | Edgar, 1955 |
| | *E. acervulina* | 4 (97 hours) | Becker, 1952; Edgar, 1955 |
| | *E. brunetti* | 5 (120 hours) | *Ibid.* |
| | *E. mivati* | 4 (93 hours) | *Ibid.* |
| | *Wenyonella gallinae* | 7–8 | Ray, 1945 |
| Pheasant | *E. phasiani* | 117 hours | Trigg, 1965 |
| | *E. meleagrimitis* | 5 (116 hours) | Clarkson, 1959a; Morgan & Hawkins, 1948 |
| | *E. meleagridis* | 5 (110–119 hours) | Edgar, 1955; Becker, 1952; Morgan & Hawkins, 1948 |
| | *E. adenoeides* | 5 (104–112 hours) | Clarkson, 1958; Moore & Brown, 1951; Edgar, 1955 |
| | *E. dispersa* | 5–6 | Hawkins, 1952; Tyzzer, 1929 |

*(Cont.)*

TABLE 2 (continued)

| Host | Species of Parasite | Prepatent Period in Days | Published Source |
|---|---|---|---|
| Pheasant—*continued* | *E. subrotunda* | 4 | Moore et al., 1954 |
| | *E. innocua* | 5 | Moore & Brown, 1952 |
| | *E. gallopavonis* | 6 | Hawkins, 1952; Moore et al., 1954 |
| Goose | *E. anseris* | 7 | Kotlán, 1932 |
| | *E. truncata* | 5–6 | Kotlán, 1933 |
| Branta canadensis canadensis | *E. fulva* | 9 | Farr, 1953 |
| | *E. hermani* | 5 | *Ibid.* |
| | *E. striata* | 5–6 | *Ibid.* |
| | *Tyzzeria parvula* | 5 | Klimeš, 1963 |
| Duck | *T. perniciosa* | 6 | Allen, 1936 |
| Pigeon | *E. labbeana* | 6–7 | Duncan, 1957 |
| Colinus virginiaus | *E. dispersa* | 5 | Tyzzer, 1929; Sneed & Jones, 1950 |

been reported in cattle and sheep coccidia.

Although the length of the prepatent period of coccidia of homoio-thermic animals is relatively stable and varies little, the coccidia of poikilothermic animals show considerable variation in the length of this period; this depends on fluctuating factors (chiefly temperature) in the external environment.

In *Barrouxia schneideri* from the intestine of *Lithobius,* the prepatent period (incubational period according to Reyer, 1937) fluctuated within significant limits. At 5° C it was 162 days; at 16° C, 50 days; at 22° C, 27 days; and at 25° C, 23 days.

Observations of *E. carpelli* have shown that when experimentally infected carp were kept in an aquarium or pond at an average tempera-ture of 16.5 to 17° C, the first oocysts appeared within 17 days after infection. If the carp were kept in water at 18° C, the prepatent period was shortened to 14 days. At a temperature of 19° C, the period of time dropped to 10 days, and at 20° C it was 7 days (Zmerzlaya, 1965). It is not clear whether the number of asexual generations was reduced or whether the existing generations simply matured faster since neither their number nor their duration have yet been established (Zaika and Kheysin, 1959).

Another species of coccidia of carp, *E. subepithelialis,* is of special interest. According to Marinček (1965), carp fry in Yugoslavia became infected with this coccidium during the spring, but development of the parasite to oocyst formation did not occur in the same season. After winter hibernation the first oocysts appeared in the yearlings during early spring at the end of March and April. Apparently, schizogony and gametogony do not occur at any time during the summer-autumn-winter period. There is some kind of latent state until March of the following year, when the cycle is completed with the formation of oocysts. Thus, the prepatent period of *E. subepithelialis* is approximately 9 months, but it consists of a period of rest and an active period, the latter being stimulated by external seasonal factors.

The fluctuation in the length of the prepatent period of coccidia of homoiothermic animals is not more than several hours. This has been observed in rabbit and chicken coccidia. However, these fluctuations increase to several days in cattle coccidia. According to Hammond (1964), the prepatent period of *E. auburnensis* fluctuated from 18 to 20 days, and according to Davis, Boughton, and Bowman (1955), from 6 to 11 days for *E. alabamensis,* and from 8 to 28 days for *E. zurnii.* I think that such large variations in time are most likely the result of faulty experimentation rather than the character of the parasite's actual development. Calves can easily become spontaneously infected right up to the beginning of the experimental infection. The investigator may interpret the early appearance of oocysts as the true length of the prepatent period, but these oocysts could have formed as a result of a premature spontaneous infection.

Small fluctuations in the length of the prepatent period depend on several causes. It was noted that the prepatent period of *E. magna* and *E. intestinalis* from rabbits, could be decreased by several hours by the use of a heavy infecting dose of oocysts as compared with a small dose (Kheysin, 1947b). When a rabbit is inoculated with one oocyst of *E. magna,* the prepatent period lasted 9 days, but a large dose reduced it to 7 or 8 days. The same was observed with *E. irresidua.* In *E. intestinalis* this difference again did not involve more than 24 to 28 hours, and in *E. media* and *E. perforans* it was no more than 8 hours. When lambs were given large doses of oocysts of *E. faurei, E. arloingi,* and *E. parva,* Krylov (1959b) observed a 1- or 2-day decrease in the prepatent period. Such changes in duration of the prepatent period depend not on a change in the length of the endogenous development, but more on an acceleration of the evacuation of newly formed oocysts from the intestine into the external environment. There is considerable damage to the intestine during an intensive infection. This leads to an early appearance of the oocysts in the feces. In addition, if many oocysts are formed, such as happens in a serious and intense case, they will appear earlier in an experimental animal. Peristalsis is also increased during an intense infection; as a result the oocysts are passed to the external environment even sooner. In a heavy infection all these factors lead to a reduction in the length of the prepatent period by several hours.

Nutrition has no influence on the length of the prepatent period. This was found in a coccidial infection of rabbits fed protein, carbohydrate, and milk diets (Kheysin, 1937b).

The length of the intestine and the place where the endogenous stages localize to some degree influence the time when the oocysts appear in the feces. The oocysts usually appear in the intestinal lumen a few hours earlier than they are found in the feces. It always takes some time for the newly formed oocysts to pass the length of the intestine. Some are even retained for a while in the cecum.

The first oocysts of *E. media* appear in the duodenum 115 to 118 hours after infection, but they do not appear in the feces until 120 to 124 hours. The oocysts of *E. irresidua,* from the middle region of the small intestine, appear in the feces after a lag of 2 to 8 hours. The oocysts of *E. magna* develop in the lower half of the small intestine by the 175th to 180th hours, but they do not appear in the feces until the 182nd to 185th hour, i.e., approximately 2 to 5 hours later. Because they form in the depths of the crypts of the large intestine and cannot be immediately eliminated from the crypts, the oocysts of *E. piriformis* also lag by 3 or 4 hours.

Gill (1954) reported that an increase in the prepatent period may depend on how long the oocysts were in the external environment prior to infection. When chicks were infected with fresh oocysts of *E. tenella, E. mitis,* and *E. maxima,* the prepatent period was normal. However, when oocysts that had been stored in 2% potassium bichromate for 2 years were used, the prepatent period lasted 2 days longer.

The length of the prepatent period may change within small limits depending on the age of the host. In year-old rabbits, the prepatent period of *E. irresidua* and *E. media* was 24 to 48 hours longer than in young rabbits. On the other hand, the prepatent period of *Isospora felis* was longer when young cats were infected than when old ones were used. In the former case it was 5 or 6 days and in the latter 3 or 4 days (Andrews, 1926, 1930).

The increased prepatent period in adult rabbits can be explained by the development of immunity which results in a suppression of certain functions of the parasite. This results in a retarded development of the asexual generations and gamogony. Henry (1932a) reported that after repeated infections of guinea pigs with oocysts of *E. caviae,* the prepatent period lengthened to 13 or 14 days instead of the 11 or 12 days of the primary infection.

Tyzzer et al. (1932) hypothesized that the length of the prepatent period of *E. necatrix* may fluctuate and depended on which kind of oocysts infected the bird. If oocysts obtained at the end of the patent period were used, the prepatent period was somewhat longer. However, if the chicks were infected with oocysts isolated during the first days of the patent period, the length of the next prepatent period was shorter. These data have not been confirmed in rabbit coccidia (Kheysin, 1947b). It is more probable that fluctuations in the length of the prepatent period are determined not by the genotypic diversity of the oocysts but by the influence of the external environment on the endogenous development of the coccidia.

Regardless of the fluctuations noted in the length of development, the length of the prepatent period for each species is constant and is determined by the genotypic qualities of the species. The different lengths of the prepatent period in various species of coccidia depend on various causes. In some cases a longer prepatent period is associated with a greater number of asexual generations preceding the onset of gametogony. For example, in the rat coccidium *E. nieschulzi,* only the fourth-generation merozoites develop into gamonts, and the prepatent period of this species is 7 or 8 days. In *E. separata* from the same host, it is the third generation which begins gametogony, and the prepatent period is 5 or 6 days. This probably explains why the prepatent period of *E. magna* is longer than that of *E. irresidua.* In the former case the third-generation merozoites begin gametogenesis in 142 hours, but in the second species it is the second generation which yields gamonts after 96 to 120 hours. The first oocysts of *E. magna* appear at the 180th hour after infection, but those of *E. irresidua* appear at the 150th hour.

In other cases it is not so much the number of asexual generations preceding the onset of gametogony which influences the length of the prepatent period, as it is the duration of schizogony. For example, in *E. media,* the third-generation merozoites, which begin gametogony, appear within 96 hours after infection, but in *E. magna* gametogony begins within 136 to 140 hours. In *E. media* the first gamonts are observed in the epithelium of the intestinal mucosa within 96 to 100 hours, and the oocysts appear in the feces within 124 hours. However, in *E. magna* the gamonts appear only within 142 hours after infection, and the prepatent period lasts 180 to 182 hours. In *E. intestinalis* the third-generation merozoites, which form the gamonts, appear on the 7th or

8th day, and the prepatent period is 10 days. In *E. bovis* the first generation of schizogony completes development within approximately 2 weeks, but the second generation is completed in 2 days. Thus, about 16 days are needed for development of the asexual generations preceding the appearance of the gamonts. The prepatent period is therefore 18 days. It is highly probable that such a long development of first-generation schizonts is determined by the large size of the schizonts and the large number of merozoites that are formed. The same is observed in *E. auburnensis* which has an 18-day prepatent period.

Apparently, *E. ellipsoidalis* does not form large schizonts. This species has schizonts with only 36 merozoites. The number of asexual generations is not known. It is assumed that the number is not large, since the prepatent period of this species is only 10 days. On the other hand, *E. zurnii* also has small schizonts, and some schizonts with only 36 merozoites have been observed; however, the prepatent period of this species is 19 days. It could be that this species has several comparatively slowly developing asexual generations.

The length of the prepatent period also depends on the length of gamete development. Gamete development takes 24 to 48 hours (in *E. coecicola*, 48 hours; *E. irresidua*, 30 hours; *E. intestinalis* and *E. magna*, 32 to 38 hours; *E. bovis*, 36 to 48 hours; *E. media*, *E. acervulina* and *E. maxima*, 20 to 22 hours). Apparently, the length of gametogony does not vary as much as the length of schizogony.

Still another factor may affect the length of the prepatent period. It has been reported that in some coccidia of chickens and rabbits, the sporozoites do not begin development in the intestine at once but only after a certain period of time. For example, in *E. tenella* this latent period in the development of sporozoites is 24 to 78 hours, and it is 48 hours in *E. magna*.

Thus, in homoiothermic animals, the length of the prepatent period of each species of coccidia is, to a certain degree, constant and comparatively immutable. The length of the prepatent period depends: (1) on which generation of merozoites begins gametogony; (2) on the length of development of each asexual generation prior to the start of gametogony; and (3) on the duration of gametogony. Several factors, which are not always present, may accelerate or retard the prepatent period. These factors apparently have little effect on the length of the asexual generations or gametogony and do not change the actual number of generations.

To date, no one has been able to experimentally alter the length of the prepatent period. This is associated with the fact that the number of asexual generations is strictly fixed, and merozoites of a particular generation simply do not form schizonts but instead begin gametogony. The number of asexual generations in the various species of *Eimeria* may vary from two to five. Two generations are found in *E. maxima, E. praecox, E. adenoeides, E. meleagridis, E. bovis,* and *E. debliecki;* three generations occur in *E. tenella, E. intestinalis, E. media, E. meleagrimitis, E. separata, E. miyairii, E. phasiani,* and, as far as is known, in *E. brunetti;* four generations are found in *E. nieschulzi* and *E. mivati;* five occur in *E. magna.* Two generations are known for *Isospora felis* (Hitchcock, 1955).

There is some question as to what factors determine the transition from asexual reproduction to gametogony. Some authors have hypothesized that this process depends on the protective mechanisms which develop in the infected host and impede continuous asexual reproduction. The resistance which the host develops after several asexual generations leads to a cessation of asexual reproduction and the start of a sexual process. Hence, the limits of a coccidial invasion are determined not by genotypic factors, but by factors in the external environment, i.e., the state of the host. This point of view has never been verified by experimentation.

Roudabush's (1935) experiments on rat coccidia and Levine's (1940) studies of chicken coccidia are highly indicative and unequivocal in this respect. Let us examine the results of the experiments on chicken coccidia. The chicks were infected with *E. tenella* oocysts. The prepatent period of this species is 7 days. On the 5th day after infection, large numbers of second-generation merozoites appeared in the ceca and then penetrated into the mucosal cells to form gamonts. After the appearance of the second-generation merozoites, 2 days passed prior to the formation of the oocysts. Hence, the 5th day in the developmental cycle of *E. tenella* was the turning point, since from this time on gametogony took place. On the 5th day after infection, merozoites were taken from the infected chicks and placed in the ceca of coccidia-free chicks. If the transition to a sexual process is determined by the state of the host itself, i.e., by the development of a resistance within the host, then second-generation merozoites transferred to a new host (uninfected by *E. tenella*) ought to continue asexual reproduction in the new host until it also develops a resistance to the parasite. It could be expected *a priori* that the resistance would arise after the same number of days as in the first host, i.e., on the 5th day. In this case it would take another 5 days after transfer of the merozoites before the sexual process could be expected to develop and another 2 days before the formation of oocysts, i.e., oocysts could be expected 7 days after the new host was infected. But the experiments showed that the oocysts appeared in the new host on the 2nd day after introduction of second-generation merozoites. The oocysts appeared at the same time as they did in the control chicks. The same results were achieved by infection of chicks with merozoites of *E. maxima, E. hagani, E. necatrix,* and *E. praecox.* In the first two species, the prepatent period was 6 days. The merozoites which develop into gamonts appeared in the intestine on the 4th day. When healthy chicks were given 4th-day merozoites, the oocysts appeared in 2 more days, i.e., in the same time as the control chicks needed to pass oocysts (the 6th day after infection by the original oocysts).

These experiments show quite clearly that the transition from asexual reproduction to gametogony is not caused by the state of the host but by genotypic factors. A particular merozoite generation has the potential to form gamonts, and it is this generation which actually establishes a new direction in the developmental cycle. Hence, even the length of the prepatent period is genotypically determined. No one has ever observed unlimited asexual reproduction in coccidia of the genus *Eimeria.*

### THE PATENT PERIOD

Two indications of the self-limiting developmental cycle in coccidia are the definite length of time during which oocysts are eliminated after a single infection and the characteristic dynamics of their elimination. It is characteristic of representatives of the genus *Eimeria* that when oocysts are used to infect a host, the elimination of the new generation of oocysts occurs after a certain period and then always ends unless a new infection has begun.

The length of the patent period fluctuates in the various species from several days to a month and even more. Among rabbit coccidia the shortest patent period is observed in *E. perforans* (6 or 7 days); in *E. intestinalis* it lasts 5 to 10 days, in *E. media,* 6 to 10 days, and in *E. coecicola,* 7 to 9 days. A longer period of oocyst elimination is observed in *E. piriformis* (up to 11 or 12 days), in *E. irresidua* (up to 16 days), and in *E. magna* (15 to 19 days). In *E. necatrix,* the patent period is approximately 12 days, in *E. phasiani,* 10 to 16 days, and in *E. mitis,* 10 days. These data were obtained in experimental infections of animals which had been carefully protected from spontaneous infection. If this condition is not met, oocysts may continue to be passed for a longer time than is observed in the absence of repeated infections, and false data about the time limit of a coccidiosis infection may be created. For example, it is highly probable that the 11- to 19-day patent period of *E. adenoeides* in turkeys (Clarkson, 1958) is somewhat overestimated, since the author remarked on the difficulty of protecting the birds against reinfection.

Apparently, it is difficult to obtain accurate data on the length of the patent period of coccidia from sheep, cattle, swine, and other animals, since protecting them against spontaneous reinfection is difficult. For a long time it seemed as if coccidiosis in these animals was not a self-limiting disease, and that the oocysts from a single inoculation could be passed with no limit. However, carefully conducted experiments demonstrated that even coccidia of cattle, sheep, and swine had a limited cycle of development which restricted the patent period to a certain length of time. For example, *E. intricata* was eliminated from sheep for only 6 to 11 days and *E. arloingi* for 8 to 10 days (Deiana and Delitala, 1953), *E. crandallis, E. faurei,* and *E. parva* for 6 to 7 days, and *E. ahsata* for 4 days (Krylov, 1959b). The oocysts of *E. auburnensis* were passed for only 2 to 7 days (Hammond et al., 1961); *E. alabamensis* for 10 to 13 days; *E. zurnii* for 3 to 28 days (average 12.8) (Davis and Bowman, 1957).

For the genus *Isospora,* the existing data on one species, *I. lacazei* from sparrows, gives no basis for speaking of a limited infection caused by this species of coccidium. The oocysts of *I. lacazei* were passed by the bird for a long time, and the infection was of a seemingly chronic nature (Boughton, 1937a). The patent period in *I. felis* was lengthy, but still the cats managed to free themselves of the infection in 27 days (Lee, 1934).

The disappearance of the oocysts from a single infection indicates the endogenous stage of development is completed. This is associated with the limited number of asexual generations.

First-generation merozoites always give rise to a new asexual generation, which causes an increase in the number of parasites in the corresponding organ of the host. The fate of the next generations of merozoites may vary. In some cases, as in *E. bovis,* the second-generation merozoites immediately become gamonts; in other species third-generation merozoites proceed to gametogony. Such is the case with *E. separata, E. miyairii,* and others. After the formation of the third-generation merozoites, which yield only gamonts, no new schizonts are formed, and as a result the infection ceases at the end of gametogony. Since the completed sexual process does not have the ability to repeat itself, the oocysts formed are passed into the external environment for a limited period of time. Schematically such a developmental cycle might be illustrated in the following manner:

$$\text{sporozoite} \rightarrow \text{sch}_1 \rightarrow \text{m}_1 \rightarrow \text{sch}_2 \rightarrow \text{m}_2$$

$$\rightarrow \text{sch}_3 \rightarrow \text{m}_3 \underset{\text{ma}}{\overset{\text{mi}}{\diagdown\diagup}} \text{zygote} \rightarrow \text{oocyst.}^*$$

It is possible that in such a scheme the number of asexual generations may increase to four or five or be only two.

In other coccidia there is a certain complication of this scheme, and second- and third-generation merozoites develop into both schizonts and gamonts. Thus, among the merozoites of one generation, there is a differentiation into potentially asexual and sexual forms. Such a phenomenon may also be observed in fourth-generation merozoites. The last generation produces only gamonts. Schematically, this life cycle can be presented as follows:

$$\text{sporozoite} \rightarrow \text{sch}_1 \rightarrow \text{m}_1 \rightarrow \text{sch}_2 \rightarrow \text{m}_2 \rightarrow$$

$$\begin{array}{l} \longrightarrow \text{ma} \\ \longrightarrow \text{mi} \end{array} \quad \text{zygote} \rightarrow \text{oocyst}$$

$$\rightarrow \text{sch}_3 \rightarrow \text{m}_3 \rightarrow \text{sch}_4 \rightarrow \text{m}_4 \overset{\text{ma}}{\underset{\text{mi}}{\diagup\diagdown}} \text{zygote} \rightarrow \text{oocyst}$$

$$\begin{array}{l} \longrightarrow \text{ma} \\ \longrightarrow \text{mi} \end{array} \quad \text{zygote} \rightarrow \text{oocyst.}$$

*Eimeria magna, E. intestinalis, E. irresidua, E. tenella,* and others have this type of development. The result is that by the time the first oocysts are formed, i.e., by the end of the prepatent period, asexual reproduction is continuing throughout the time required for completion of the following asexual generation. Therefore, at the beginning of the patent period, asexual generations still remain in the epithelium of the

---

*Sch = schizont; m = merozoite; mi = microgametocyte; ma = macrogamete.

intestine and continue development. As a result the patent period is extended and always appears longer than in species in which asexual reproduction ends by the time the first oocysts appear. One way or another, endogenous development always ends and continued production of oocysts cannot be accomplished, and the patent period is limited in time.

The extension of the patent period may depend on asynchronous development, mainly of the asexual generations. It has been observed that in chicken and rabbit coccidia, the sporozoites do not all begin development at the same time, so that the end of the first generation of schizogony may sometimes extend more than 24 hours. Therefore, the second generation of *E. tenella* does not start simultaneously but may be between 96 and 120 hours after the primary infection. In rabbit coccidia such a lag is also present and results in the failure of some parasites to penetrate the tissues of the intestine on schedule. Sometimes in *E. magna,* asynchronous development of sporozoites, and likewise the first asexual generation, causes second- and third-generation schizonts to be present in the rabbits' intestine for 3 or 4 days. This is also expressed in the duration of the patent period.

If all the merozoites of one generation enter into gametogony and if endogenous development ends at approximately the same time (or somewhat later than the first oocysts appear) the patent period will be comparatively short. For example, in *E. miyairii* and *E. separata,* the prepatent period lasts 5 to 8 days. Whenever only part of the merozoites of a given generation begin gametogony and endogenous development exceeds the duration of the prepatent period, the patent period is extended considerably because oocysts form over a longer span of time. An example is found in *E. magna, E. irresidua,* and *E. piriformis,* in which the patent period is 15 to 19, 16, and 12 days, respectively. The patent period may sometimes be extended as a result of oocysts being retained in the intestinal tissues for some time after endogenous development has ended. This phenomenon has been observed in chickens infected with *E. tenella.* The oocysts of this species are in the walls of the ceca for several months, and are passed into the external environment in small numbers (Herrick, Ott, and Holmes, 1936). The same has been observed for *E. piriformis, E. coecicola,* and *E. magna* from the rabbit. Packets of oocysts are sometimes observed in the intestinal mucosa as long as a month after the end of endogenous development.

The elimination of oocysts is characteristic during the patent period of a single infection.

## DYNAMICS OF OOCYST OUTPUT DURING THE PATENT PERIOD

The number of oocysts eliminated is comparatively low on the 1st day of the patent period, but on the 2nd and 3rd day the maximum is reached, and on subsequent days fewer are found until none are present in the feces (Kheysin, 1947b; Hammond et al., 1963). Fig. 43 illustrates a rather typical curve of the dynamics of oocyst abundance during the patent period of a rabbit infected with *E. magna*. Depending on the degree of synchronism in the development of the asexual generations, the peak of oocyst output may be very limited in time or very lengthy. In *E. perforans* and *E. intestinalis* infections, many oocysts are passed on the 2nd or 3rd day, but then the numbers drop off sharply. In *E. magna,* the peak of oocyst output is extended over 3 or 4 days.

When rats were infected with oocysts of *E. miyairii* the following dynamics of elimination were observed during the patent period:

| | |
|---|---|
| 1st day | $1.73 \times 10^4$ |
| 2nd day | $92.56 \times 10^4$ |
| 3rd day | $206.23 \times 10^4$ |
| 4th day | $146.00 \times 10^4$ |
| 5th day | $64.98 \times 10^4$ |
| 6th day | $21.69 \times 10^4$ |
| 7th day | $4.71 \times 10^4$ |
| 8th day | $0.64 \times 10^4$ |
| 9th day | $0.06 \times 10^4$ |

On the 1st day 0.32% of all oocysts were passed, and on the days following, 17.18, 38.28, 27.11, 12.06, 4.03, 0.78, 0.12, and 0.11%, respectively (Becker, Hall, and Hager, 1932).

When synchronism is high and there are few asexual generations, there is an ever greater concentration of oocysts at one time. *Eimeria bovis,* which has two generations of schizonts, delivers most of the oocysts on the 2nd and 3rd day of the patent period. *Eimeria intestinalis* has three asexual generations, with the second being the most productive, as a result of which about 50% of the oocysts are passed during the first 4 days. *Eimeria magna* has five generations, and a relatively large percentage of the oocysts are passed during 10 days.

If oocyst output lasts long and the number of oocysts begins to increase by the end of the patent period, the possibility of a repeated infection occurs, and it is assumed that such took place during the pre-

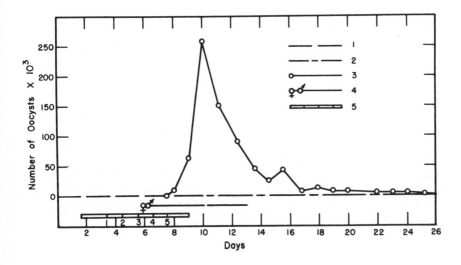

Fɪɢ. 43. Abundance of *E. magna* oocysts eliminated during the patent period (After Kheysin, 1940). *1*, Prepatent period; *2*, patent period; *3*, number of oocysts passed; *4*, gametogony; *5*, asexual reproduction.

patent period. This is easily seen in Fish's (1931a and b) data on the output dynamics of *E. tenella* in one chick:

| Days after infection | No. of oocysts passed |
| --- | --- |
| 1–7 | 0 |
| 8 | 168,000 |
| 9 | 60,000 |
| 10 | 3,000 |
| 11 | 212,500 |
| 12 | 2,500 |
| 13 | 37,700 |
| 14 | 25,600 |
| 15 | 197,400 |
| 16 | 235,600 |
| 17 | 68,600 |
| 18 | 2,200 |
| 19–20 | 0 |

It was probably for that very reason that the chicks which were infected with *E. tenella* had two peaks of oocyst output on the 4th and 9th day of the patent period. As shown in Fish's experiments, the sharp drop in the number of oocysts eliminated on the 17th to 19th days after infection can be explained by a newly developed, superinfective, acquired immunity which exerted a limiting effect on the endogenous stages of the parasite.

Therefore, one can definitely speak of the infection as being self-limiting for various species of the genus *Eimeria;* as a result of the single inoculation, the phenomenon is expressed externally as a gradual disappearance of the oocysts from the feces.

It is a bit more complicated with representatives of the genus *Isospora.* Sparrows and some of the other passerines are often infected with *Isospora lacazei.* This coccidium is apparently cosmopolitan and can be found wherever the proper host exists. According to the observations of many investigators (Boughton, 1929, 1930, 1937a; Scholtyseck, 1954, and others), sparrows are very often severely infected with *Isospora,* and it is difficult to find birds which are coccidia-free. The birds are usually infected by *Isospora* in the nest. Observations by Boughton (1937a) showed that the sparrows in a laboratory continued to discharge considerable numbers of oocysts for several months, even though they were originally infected only once and were kept in cages with a mesh floor, which allowed a minimum possibility of self recontamination. Below are the figures on elimination of oocysts of *Isospora lacazei* (in thousand per bird per day) in two groups of sparrows (Boughton, 1937a):

| March | Group 1 | Group 2 | April | Group 1 | Group 2 |
|-------|---------|---------|-------|---------|---------|
| 1 | 9[1] | 23 | 1 | 199 | 12 |
| 2 | 8 | 17 | 2 | 128 | 12 |
| 3 | 9 | 24 | 10 | 92 | 28 |
| 4 | 20 | 11 | 11 | 72 | 18 |
| 5 | 9 | 7 | 12 | — | 48 |
| 6 | 5 | 18 | 16 | — | 93 |
| 7 | 11 | 17 | 18 | — | 60 |
| 8 | 7 | 21 | 19 | 173 | 41 |
| 9 | 4 | 21 | 21 | 67 | 19 |
| 10 | 29 | 16 | 22 | 149 | 49 |
| 11 | 415 | 20 | 23 | 183 | 54 |
| 12 | — | — | 24 | 101 | 62 |
| 19 | 150 | 12 | 25 | 97 | 32 |
| 21 | — | — | 26 | 89 | 48 |
| 25 | 375 | 28 | 27 | 109 | 33 |
| 26 | 221 | 26 | 28 | 81 | 51 |
| | | | 29 | 68 | 61 |
| | | | 30 | 88 | 56 |

[1] A supplementary experimental infection of sparrows with oocysts of *I. lacazei.*

From these data, it is apparent that in the second group the number of oocysts remained at the same level for approximately 2 months, while in the first group there were small fluctuations and a large number of oocysts were passed over a considerable period of time. Boughton (1937a) hypothesized that in *Isospora* the infection is not self-limiting in time and is of a chronic nature. The possibility has not been eliminated that there was a constant reinfection or superinfection, with insufficient acquired immunity to protect against the new invasion. In any case, this question still remains unanswered.

An analogous phenomenon of long term elimination of oocysts has been observed during infections of carp by *E. carpelli*. The young fish were infected in breeding or growing ponds in early spring, and they remained carriers of coccidia throughout the summer and autumn. Oocysts were seen in their feces every day for several months, and the impression of a chronic infection was created. Experiments designed to resolve this question yielded the following results. Two-month-old coc-cidia-infected carp were transferred to aquaria. Some of the aquaria were cleaned three times a day, and the water was completely changed. In the control aquarium, no water change was made. The fish in the control aquarium remained infected with coccidia for 2 months. The carp in the experimental aquaria were completely free of coccidia after 1 month (Zmerzlaya, 1965). Although it was not simple to establish the fact, *E. carpelli* has a self-limiting infection. Under natural conditions the fish are constantly reinfected, and with no immunity the endogenous stages can repeat themselves endlessly. The so-called chronic invasion of carp is therefore caused by constantly repeated infections. It is highly probable that the same phenomenon occurs when birds are infected with *Isospora*.

## DAILY DYNAMICS OF OOCYST OUTPUT

In a study of *Isospora* in sparrows, Boughton (1929, 1930, 1933, 1934, 1937a, and b) found the interesting fact that the oocysts are eliminated in the feces at certain times of the day. By studying sparrow droppings infected with *Isospora* every 6 hours, Boughton found that during the period from 9 P.M. to 9 A.M. no oocysts are eliminated. An insignificant number of oocysts was found in the droppings eliminated between 9 A.M. and 3 P.M. For example, on the 10th day after infection a few birds passed an average of 9000 oocysts daily per bird, but the greatest number of oocysts was passed in the period from 3 P.M. to 9 P.M. In the same experiment, each bird yielded an average of 406,000 oocysts during this period. It was observed that the evening peak of oocyst elimination did not depend on the time when the birds were infected. Whether the original infection began in the morning or the evening, the maximal number of oocysts still appeared between 3 P.M. and 9 P.M. Schwalbach (1961) observed an analogous rhythm of oocyst output in *I. lacazei* from sparrows. The same periodicity of oocyst output has been observed in other species of *Isospora* from various passerine birds studied in zoological gardens. Approximately 80% of the daily number of oocysts were passed between 3 P.M. and 9 P.M.

Periodicity has also been observed in the infection of birds by several species of coccidia of the genus *Eimeria*. Boughton (1937b) observed that most oocysts of *E. labbeana* from pigeons are passed between 9 A.M. and 3 P.M., in lesser amounts between 3 P.M. and 9 P.M., and even fewer emerge during the nighttime hours. For example, in a study of five pigeons over a 7-day period following their infection by *E. labbeana*, it was established that an average of 23,627 oocysts emerged from 9 A.M. to 3 P.M. From 3 P.M. to 9 P.M., 4832 oocysts were evacuated, and from 9 P.M. to 9 A.M., 1168. During the daytime hours approximately eight to ten times more oocysts were eliminated than during the evening hours. It was sometimes impossible to find any oocysts in the night feces, but they were easily found during the day.

By comparison of these data with the periodicity of oocyst elimination in *Isospora*, it is evident that there is no correlation in time of maximal oocyst elimination between *I. lacazei* and *E. labbeana*. In the latter species, the peak of oocyst output was approximately 6 hours earlier than in *I. lacazei* from sparrows.

Levine (1942a and b) observed a periodicity of oocyst elimination in five species of *Eimeria* from chickens. This phenomenon was especially pronounced in *E. hagani*. In an experimental infection, the oocysts of this species were chiefly eliminated from 7-week-old chicks between 3 P.M. and 9 P.M. During this time, 87 to 91% of all daily oocyst output took place. During these same hours, the maximum number of *E. praecox* and *E. mitis* oocysts were also eliminated. However, as compared with *E. hagani*, 60% of all oocysts of *E. praecox* and only 50% of *E. mitis* were eliminated between 3 P.M. and 9 P.M. The remaining oocysts were passed over 18 hours from 9 P.M. to 3 P.M. the next day, with no particular peak during this period. The oocysts of *E. brunetti* were primarily passed from 9 A.M. to 3 P.M. (50 to 60%). Up to 19% of the oocysts were discharged from 3 P.M. to 9 P.M., and 26 to 37% from 9

P.M. to 9 A.M. The oocysts of *E. maxima* have a unique schedule. The peak output has no special preference for the time of day or night. In one experiment lasting 3 days, the number of oocysts was greatest between 3 P.M. and 9 P.M., but on 2 other days the maximum was from 9 A.M. to 3 P.M. The lowest number of oocysts was passed from 9 P.M. to 9 A.M. A different daily rhythm was found in *E. necatrix*. The greatest number of oocysts (47 to 62%) of this species was passed at night and early in the morning during the period between 9 P.M. and 9 A.M. No clear periodicity of oocyst output has been observed for *E. tenella.*

No daily rhythm of excretion has been observed for rabbit coccidia, and apparently the coccidia of cattle and sheep have no output rhythm either. Daily periodicity is thus observed only in coccidia of birds, and even so not among all bird coccidia. Apparently, this can be explained by the fact that birds show the greatest rhythm in their daily physiological activities. From approximately 6 A.M. to 6 or 7 P.M., many birds lead an active life, move around a great deal and feed, but during the remainder of the day they rest and take no food. During the daylight hours they eliminate the greatest amount of feces. One would think that the periodic elimination of oocysts among many coccidia is to a great degree associated with the physiological processes of the host itself.

The maximum number of *Isospora* oocysts is excreted from sparrows which are on a normal day-night regime at the end of the light period (approximately 6 P.M.). The physiological activity of the birds during the entire light period aids in the process of oocyst formation, the freeing of oocysts from the cells of the intestinal mucosa, and their elimination from the body. This process lasts no less than 12 hours. Immediately after the rest period oocyst formation starts again, and by the end of the light period, a massive release of the oocysts and elimination with the feces begins. During the night hours when the bird is not active, the sexual process apparently slows down because of the influence of certain external factors, as a result of which the formation and elimination of oocysts does not occur.

The normal periodicity of oocyst elimination in coccidia of birds has been observed only in those species which are located in the small intestine. The oocysts from the small intestine freely enter the feces. In those species which are located in the ceca (e.g., *E. tenella* and *E. necatrix*), periodicity of elimination is somewhat different from that of species which live in the small intestine. Of the oocysts of *E. necatrix*, 47 to 62% were passed between 9 P.M. and 9 A.M. Apparently, emptying of the ceca occurs independently of the rest of the intestine, and the cecal contents may enter the cloaca separate from the contents of the small intestine (Ikeda, 1956b). Therefore, the elimination of the oocysts of *E. necatrix,* like those of *E. tenella,* does not have as distinct a periodicity as other species of coccidia from the small intestine.

The relationship between oocyst elimination and the physiological activity of birds has been well demonstrated by experiments which alter the activity regime of birds during a 24-hour period. If infected sparrows are placed in surroundings which create complete artificial illumination during the night (from 6–7 P.M. to 8–9 A.M.) and darkness during the daytime (artificial night), within a few days the dynamics of oocyst elimination will be greatly altered. Under such reversed conditions, the birds sleep during the day and move around and feed during the actual night. Within 4 days the maximum oocyst elimination shifts from 6 P.M., as is observed during normal hours, to 6 to 8 A.M., i.e., to the end of the artificial light period. During the first 3 days after the birds are placed in the new surroundings, the maximum output gradually shifts to the morning hours. When the normal regime is restored, the rhythm of elimination also returns to normal.

If birds are kept in darkness from 12 midnight to 12 noon and in light from noon to midnight the maximum oocyst elimination is at the end of the light period, i.e., from 11 P.M. until 2 A.M. Hence, elimination of oocysts is associated with some kind of physiological process which is completed during the active period of the bird's daily life. It is possible that this is determined by the muscular activity of the intestine or by feeding which occurs only during the light period. However, if a bird is not fed between 6 A.M. and 12 noon during a normal day-night regime, the maximum oocyst output remains unchanged and occurs at 6 P.M., i.e., at the end of the light period. The same is observed when the birds are left hungry from 12 noon to 6 P.M. Thus, it seems that food does not have any direct effect on the rhythm of oocyst output.

The minimum time necessary for development of gametes and the formation of oocysts in *I. lacazei* is apparently 24 hours. So far, no one has been able to accelerate this process by altering the physiological state of the host. This was attempted in an experiment during which the birds were confined and subjected to alternate light and dark periods of 6 hours each. In some experiments, light was administered in two periods from 6 A.M. until noon and from 6 P.M. until midnight; in other experiments there was light from 12 noon until 6 P.M. and from midnight to 6 A.M. In the first experiment the maximum number of oocysts came only between 6 P.M. and 10 P.M., i.e., after the second 6-hour period of darkness. It might be expected that there would be two peaks of oocyst elimination at the end of each light period, i.e., every 12 hours. But the development of gametes and the formation of oocysts could not be speeded up, and the 24-hour scheme of development remained with only one peak of oocyst elimination per day. In the second experiment the maximum number of oocysts was passed 6 hours earlier than in the first, i.e., at approximately 12 noon.

In order to determine how the length of the light period affects the periodicity of oocyst elimination, Boughton designed a series of experiments with sparrows that had been placed either in short day surroundings with the 24-hour cycle reduced to 19 hours, or in long day conditions with a 24-hour day extended to 29 hours. In the short day chambers, the birds had 10 hours of light and 9 hours of darkness. The other group had 16 hours of daylight and 13 hours of darkness. The sparrows with short days passed oocysts of *I. lacazei* irregularly with the greatest number arriving at the start or end of the light period. The intervals between peaks were approximately 18 hours. Elimination of oocysts from sparrows with long days occurred mainly during the light period, with 30 hours between peaks. In both series of experiments the peak of oocyst elimination was not timed to any particular part of the day but always coincided with the artificial period. Thus, the length of the sexual process and the period of oocyst formation, as well as the period when the oocysts were eliminated from the cells changed somewhat under artificial conditions of day length.

All these data reveal that the process of preparing the oocysts for elimination from the intestinal epithelial cells is, to a significant degree, controlled by the physiological state of the host. Probably some periodical changes in the metabolism of the host, associated with a change in light and darkness, affect the intracellular development of the coccidia and lead to a daily periodicity in oocyst elimination.

## NUMBER OF OOCYSTS ELIMINATED
## DURING THE PATENT PERIOD

Each coccidian species is characterized by the formation of a certain number of oocysts. This depends on several factors: first, the number of asexual generations and the number of merozoites produced by each generation, as well as the number of merozoites which go on to become macrogametes; second, the external conditions which may influence the survival of the merozoites and other stages of development of the parasite and a decrease or increase in the number of oocysts. The first of these factors is determined by the parasite itself; the second is determined by the host.

Each oocyst of the genus *Eimeria* or *Isospora* contains eight sporozoites; this must be accounted for in considering the productivity of the parasite. If, for example, one theoretically computes the possible production of any species of coccidium on the basis of what one oocyst can produce, the following data are obtained. *E. magna* has five generations of merozoites. Each generation produces an average of 10 to 30 merozoites. Some of the third- and fourth-generation merozoites form gamonts, and fifth-generation merozoites all form gamonts. The numerical ratio of the merozoites with differing sexual potential is not known, so it is difficult to determine exactly what percentage of the merozoites will yield macrogametes and microgametocytes. If it is assumed that half of all merozoites form macrogametes, then a rabbit infected with one *E. magna* oocyst should produce about three million oocysts. However, only about 800,000 oocysts are actually passed (Kheysin, 1940, 1947b). The actual number of oocysts produced depends on the number of merozoites that die (apparently during the period they are moving in the intestine). The productivity of *E. intestinalis* is somewhat greater; 1,200,000 oocysts can be produced from one original oocyst. This species has only three generations of merozoites, but each forms an average of 50 to 60 merozoites. *Eimeria media* forms about 150,000 oocysts per original, but *E. coecicola* produces 100,000 (Kheysin, 1947b).

Roudabush (1937) calculated the number of merozoites in each generation and found that theoretically, one oocyst of *E. nieschulzi* could form an average of 1,872,000 gamonts. When a rat was infected with six oocysts, a maximum number of 1,455,000 oocysts was obtained for each original oocyst (Hall, 1934). The author hypothesized that the difference of 417,000 between the theoretical number of gamonts and the number of oocysts is attributable to the number of microgametocytes formed from one oocyst. *Eimeria miyairii* and *E. separata* should theoretically yield an average of 38,016 and 1,536 gamonts per inoculated oocyst. The difference in numbers of gamonts formed among the three species, and likewise the number of oocysts, is associated with the different number of asexual generations and the varying productivity of the schizonts of each generation. *Eimeria nieschulzi* forms four generations of merozoites, but the two other species form three generations, each with a small number of merozoites. This is expressed in the total number of oocysts passed by these species. When rats were given doses of 1,500 *E. miyairii* oocysts every day for 5 days, they passed an average of 5.4213

$\times$ $10^7$ oocysts throughout the patent period. When infected with *E. separata,* they yielded 2.459 $\times$ $10^6$ oocysts (Becker et al., 1932).

*Eimeria tenella* from chickens, which has three generations of merozoites, should theoretically pass about 2,520,000 oocysts, but in reality the administration of one oocyst will yield about 400,000. *Eimeria brunetti* has a similar productivity per oocyst. It forms a maximum of 400,000 oocysts. In other species of coccidia from birds, the productivity is considerably lower. *Eimeria acervulina* develops a maximum of 72,000, *E. necatrix* develops 58,000, and *E. maxima* 12,000 oocysts (Brackett and Bliznick, 1952). However, these figures may vary from one experiment to another and will change as the number of oocysts in the original infecting dose is changed.

The number of oocysts capable of being eliminated in the primary infection of a nonimmune host depends to a large degree on the dose which was originally given to the host. When the inoculating dosage is increased, a proportional increase in oocyst output is not observed. With comparatively minor dosages, proportionality is preserved, but if the infective dosage is greatly increased it will not lead to an unusual increase in the number of oocysts passed. On the contrary, the number of oocysts eliminated per original oocyst may actually decline. For example, when each rat was given one *E. nieschulzi* oocyst, 62,000 oocysts were passed per rat. When a single rat was inoculated with six oocysts, the total number of oocysts eliminated increased to 1,455,000 per original single oocyst. Later, when rats were inoculated with 150 oocysts only 1,029,666 oocysts emerged per original oocyst. Finally, with an infective dosage of 2,000 oocysts, only a total of 144,150 were excreted (Hall, 1934; Roudabush, 1937).

When each 2- and 3-week-old chick was infected with 200 *E. necatrix* oocysts, approximately 50,000 oocysts per original oocyst were eliminated. In chicks given 2,000 oocysts, only 2,500 oocysts per original were recovered. The same phenomenon has been observed in experiments with 10-day-old chicks. When each was given 50 oocysts, each chick yielded 3 million oocysts, or approximately 58,000 oocysts per inoculative oocyst. When each was given 250 oocysts, the yield was 20,000 oocysts per inoculative oocyst, and with infecting doses of 1,250, 6,250, and 350,000, the oocyst yield was 2,000, 1,000, and 100 oocysts per original oocyst, respectively.

Similar results have been obtained with *E. tenella*. When 2-week-old chicks were inoculated with 50 and 250 oocysts, the yield per original oocyst was 80,000 and 60,000 respectively. When the original dose was 6,250 oocysts, a total of 10,000 [per oocyst] were recovered, but administering 20,000 and 40,000 resulted in 1,750 and 1,200 oocysts [per oocyst], respectively.

In different experiments when 2- and 3-week-old chicks were given 50 oocysts of *E. brunetti* each, the harvest was from 108,000 to 400,000 oocysts [per oocyst], but an infecting dose of 1,250 oocysts produced 24,000 to 26,000 oocysts [per oocyst], and when given 20,000 and 40,000 oocysts, the corresponding numbers were 800 to 1,700 and 400 to 500 oocysts (Brackett and Bliznick, 1952). When chicks were infected with 2,000 oocysts of *E. acervulina,* the yield was from 35,000 to 72,000 oocysts per original. An inoculation of 20,000 oocysts produced only 7,600 oocysts per original (Brackett and Bliznick, 1949).

This principle was revealed even more clearly by Krassner (1963) with infections of 3-month-old chicks with various doses of oocysts of *E. acervulina.* A dose of 1,000 oocysts led to the elimination of 6,590 oocysts per original. At higher doses of $5 \times 10^3$, $10^4$, $10^5$ and $10^6$ oocysts, 6,180, 4,150, 444, and 70.8 oocysts were eliminated, respectively. Chicks infected with 200 oocysts of *E. maxima* yielded 11,500 oocysts per original oocyst, while an infecting dose of 2,000 produced only 2,250 oocysts per original. In various experiments, infection with 10,000 oocysts yielded only 940 to 2,900 oocysts.

It is apparent that when the dosage is increased, the productivity of the parasite drops in varying degrees for each species. For example, with *E. tenella,* 50 oocysts yield as many as 4.4 million, or 80,000 oocysts per original inoculated oocyst, but when the infective dose is 40,000, the average bird yields 47.9 million oocysts, i.e., 1,200 oocysts per original oocyst. An infective dose of 40 *E. brunetti* oocysts yields 5.4 million oocysts, or 108,000 per original oocyst, and an infecting dose of 40,000 yields 21.6 million, i.e., 500 per original.

It has also been observed that the maximum relative number of oocysts (number per original oocyst) in the various species is obtained when birds are inoculated with different numbers of oocysts. Thus, in *E. necatrix,* the greatest number of oocysts (58,000) is passed when the infecting dose is 50 oocysts, but in *E. acervulina* it is achieved with a dose of 2,000 (up to 72,000 oocysts). The maximum absolute number of oocysts eliminated is also different in the different species. For example, *Eimeria acervulina* is most productive when the infecting dose in one bird is 50,000. An average of 430 million oocysts is collected from each bird. The maximum number of *E. tenella* and *E. brunetti* oocysts occurs after an infecting dose of 6,250 oocysts. The average yields per bird at this dosage are 65 and 53 million, respectively. The maximum number of *E. maxima* oocysts (36 million) is obtained when the original dose is

10,000. The least productive is *E. necatrix.* When a bird is infected with an initial dose of 20,000, the yield is 12 million oocysts which is a maximum for this species (Brackett and Bliznick, 1952).

The decrease in productivity of oocysts at high dosage levels is apparently associated with "overpopulation" of the endogenous stages in the epithelium of the intestinal mucosa, as a result of which, there is a high death rate not only of the merozoites, but apparently also of the other stages of development. In a heavy infection of chicks with various species of coccidia, especially *E. tenella,* a multitude of free merozoites, which must have ultimately died, were observed in the bloody feces passed on the 5th and 6th days after infection (Tyzzer et al., 1932; Brackett and Bliznick, 1952). As a result, the productivity of the oocysts decreased when the infective dose was increased. Apparently, the effect of overpopulation is different in the various species. It is less pronounced in *E. acervulina,* in which the shortest developmental cycle is found, and strongest in *E. necatrix* and *E. tenella* which have a longer cycle.

It is also possible to assume that the decrease in parasite productivity following high infective doses is associated with the fact that the very first asexual generations cause the host to develop a rapid immunity which limits the development of further schizonts and slows down gametogenesis. At low doses this effect is not so pronounced since a smaller amount of antigen is produced; however, this hypothesis requires confirmation.

All the above examples concern cases of primary invasion by a given species of coccidium. The productivity of coccidia may drop sharply when a relatively immune host is invaded. In a repeat infection of rabbits with oocysts of *E. intestinalis,* the productivity of the parasite decreased three times as compared with the original dose. During the third infection of the rabbit with a dose of 50,000 oocysts of *E. intestinalis,* which followed two infections with 5,000 and 10,000 oocysts, the productivity of the parasites dropped 20 times per single oocyst.

The older the animal, the lower the number of oocysts produced per original. For example, in 9-day-old chicks given a dose of 200 oocysts of *E. necatrix,* an average of 5.15 million oocysts emerged (26,000 per original oocyst). The same dose in 2-month-old-chicks gave different results; each infected bird yielded 2.5 million oocysts (12,500 per original). There was a 50% drop in the productivity of the oocysts (Brackett and Bliznick, 1952). This is due to the increasing immunity of older birds.

The possibility has not been eliminated that a factor such as the host's feeding habit may exert a certain influence on the numbers of oocysts

formed. In this respect the data obtained during an infection of rats by
*E. nieschulzi (E. miyairii)* are interesting (Becker and Morehouse, 1937).
The number of oocysts passed increased if the diet contained yeast as a
rich source of vitamin $B_2$. The inoculating dose in all cases was the
same. On a diet containing 10% yeast, an average of 207,670,000
oocysts were eliminated; if there was 6% yeast in the diet, 172 million
oocysts emerged during the patent period, but on a standard diet the
yield was only 137,670,000. If the rat diet contained fresh liver or a dry
liver powder considerably fewer oocysts were produced (in various
experiments from 39.56 to 101.4 million). If the rats were given dry milk,
very few oocysts were obtained (about 8 million). The addition of dry
milk to the standard diet containing polished rice (1.5%) led to the elimination of an average of 24,710,000 oocysts, while the addition of 7% dry
yeast to the same diet brought about an increase to 66 million oocysts.
On a standard diet with 6% rice, an average of 68 million oocysts were
produced. Thus, a drop in the amount of vitamin $B_2$ in the diet of the
animal exerts an influence on the endogenous stages, the final result of
which is expressed in the number of oocysts eliminated. Feeding rabbits
milk, protein, and vegetable food did not lead to a change in the productivity of *E. magna* and *E. stiedae*.

# CHAPTER V

# Sporulation of Oocysts and Their Survival in The External Environment

As was noted earlier, in most species of coccidia the oocysts emerge from the host into the external environment in an undeveloped state, and only under certain conditions does the process of sporogony or sporulation begin. The basic conditions for sporulation in the external environment are a suitable temperature, sufficient moisture, and free access to oxygen. It is highly probable that the oocysts of many species cannot begin sporulation in the intestine of the host because of a permanent oxygen deficit, and in warm-blooded animals the temperature in the intestine is too high for the normal process of sporogony. A few species of coccidia from poikilothermic animals (fish, reptiles) and *Isospora bigemina,* from the intestine of the dog, are the exceptions. Despite the oxygen deficit, sporulation occurs in the intestine. Apparently, the small amount of oxygen present in the intestinal tissues is sufficient for this process. The sporulated oocysts emerging from the host into the external environment immediately become infective and need no further maturation.

It is a different matter when unsporulated oocysts are excreted into the external environment. They need time and favorable environmental conditions to mature. Light has no effect on sporulation, and the oocysts can develop both in darkness and in light (Pérard, 1925; Kheysin, 1937a; Duncan, 1959a).

## SIGNIFICANCE OF OXYGEN DURING SPORULATION

Sporulation cannot occur in an environment lacking oxygen. Any factor which removes oxygen from the vicinity of the oocysts will impede spo-

149

rogony. If the feces containing the oocysts are placed in water, decomposition will begin after a certain period of time. Here, sporulation of the oocysts is retarded because of a lack of oxygen; the large numbers of bacteria developing in such an environment will compete with the oocysts for oxygen. If the bacteria are removed from the environment in which the oocysts are found, sporulation will take place. Because of the negative effect of bacteria on oocyst sporulation, cultivation of the oocysts is always done in a medium from which putrefaction has been eliminated and favorable conditions for aeration have been created. The best medium for culturing oocysts is a 2% solution of potassium bichromate ($K_2Cr_2O_7$). In this solution the oocysts have complete access to oxygen, and bacteria do not usually develop. Any other solution which suppresses putrefaction may be used for culturing oocysts. In $K_2Cr_2O_7$ the oocysts cannot only sporulate but can be preserved alive for long periods of time. This solution does not penetrate the wall of the oocyst and will not kill the zygote or the sporozoites. The oocysts of rabbit coccidia were viable after 2 years in potassium bichromate (Kheysin, 1959a). The same has been observed for chicken and livestock coccidia. Oocysts of most species of coccidia have survived more than 2 years, but all have died after 3 years in this medium (Gill, 1954). Oocysts of *E. zurnii* have survived for the same length of time (Petrov and Nikonov, 1964). It is only necessary to add some kind of antibiotic to the water, and the drug will aid the development of the oocysts by inhibiting bacterial growth. Sporulation of oocysts washed free of feces is possible and occurs in pure water without the addition of antibiotics. For this effect, it is necessary to clean the oocysts very carefully so that the absolute minimum number of bacteria get into the water. The development of oocysts in water can be accomplished if the water is continually aerated. In such cases, oocysts even sporulate in containers with large amounts of water. If the water is not aerated, the oocysts at the bottom of a deep container of water sporulate poorly and slowly.

A lack of oxygen impedes the primary phase of sporogony. Oocysts with spherical zygotes do not begin sporulation under anaerobic conditions, and division of the nuclei does not occur. After the oocysts are given free access to oxygen, sporulation begins and development of the oocysts proceeds normally.

As was described above, during sporulation of *Eimeria* oocysts, the first thing observed was the formation of four outgrowths on the surface of the spherical zygote; then the four spherical sporoblasts separated, and the pyramid stage was formed. After the pyramid stage the round sporoblasts formed again and stretched to become oval-shaped. At this time the wall appeared on their surfaces, and the sporoblasts developed into sporocysts. All these changes occurred without division of the nuclei (as was shown for *E. magna* and *E. stiedae*) (Kheysin, 1935b). All such morphological changes, not accompanied by nuclear division, can occur under anaerobic conditions. If the oocysts which have reached the stage with four round sporoblasts or the pyramid stage are placed in an anaerobic environment within 3 days, development will have proceeded so that oocysts with oval sporoblasts or even sporocysts are present. However, completion of sporulation does not occur because one more division of the nuclei has yet to take place in the sporocysts so that sporozoites are formed; this division cannot occur under anaerobic conditions.

Within 24 hours at room temperature and free access to oxygen, 18.5% of the oocysts of *E. magna* and *E. stiedae* were at the four-sporoblast stage, 28.3% were in the pyramid stage, 5.9% were in the round-sporoblast stage, 2.2% were in the oval sporoblast stage, and 45% had not begun to develop. After 72 hours under anaerobic conditions, the number of oocysts with four sporoblasts or pyramids had dropped to 3.0 and 1.4% respectively. Round sporoblasts accounted for 0.9%, but the number of sporocysts had risen to 56.5%. Hence, anaerobiosis did not impede morphological reorganization during sporogony (Kheysin, 1935b).

Goodrich (1944) reported that coccidian oocysts can develop in an oxygen deficit. However, it is unclear whether they reach the infective stage, or whether they accomplish the above described changes without nuclear division. Duncan (1959a) mentioned the possibility of sporulation of oocysts of *E. labbeana* (the pigeon parasite) under conditions of oxygen deficit, but only a few oocysts sporulated.

Oocysts of *E. zurnii* sporulated only when the oxygen deficit was 10% or less. If the deficit was greater, the time of sporulation was increased, and in the complete absence of oxygen no development took place (Marquardt, Senger, and Seghetti, 1960). The development of several species of oocysts in the tissues of the host is possible only when these oocysts have a weak demand for free oxygen.

## SURVIVAL TIME OF OOCYSTS UNDER ANAEROBIC CONDITIONS

Metzner (1903) placed unsporulated oocysts of *E. stiedae* in anaerobic surroundings for 2 to 3 weeks, after which he checked their viability. All the oocysts had died. An experiment was performed to determine the ability of the oocysts to develop in potassium bichromate under aerobic conditions.

According to Kheysin (1935a), the unsporulated oocysts of *E. magna, E. stiedae,* and *E. perforans* completely re-established development after 5 or 6 days in anaerobic surroundings. However, 3 weeks in such an environment led to the death of 90% of the oocysts. After 3 to 5 days in anaerobic surroundings, unsporulated oocysts of *E. labbeana* remained viable, but after 10 days of exposure, all died (Duncan, 1959a).

Placing oocysts of *E. magna* in bouillon which contained a high concentration of bacteria led to a 100% mortality of the oocysts by the 5th day. No oocysts developed after being transferred to bichromate solution. Oocysts of *I. felis* and *I. rivolta* developed at 16 C in bouillon and were able to survive for 12 to 15 days. At 36.5 C isolated oocysts (up to 6%) appeared in the putrefactive environment within 72 hours; the oocysts had two sporoblasts. Within 10 days, only 10% of the oocysts sporulated after being placed in a 2% potassium bichromate solution. It is possible that in this case the death of the oocysts resulted not only from an oxygen deficit, but more likely as a result of the toxic effects of by-products produced by the bacteria. One way or another, long term confinement of oocysts in a putrefactive environment caused their death, and sporulated oocysts lost their ability to infect animals.

Oxygen is used by the oocysts for respiration. Wilson and Fairbairn (1961) found that the intensity of oxygen usage by oocysts of *E. tenella* was much higher in newly formed, unsporulated oocysts than in sporulated ones. The intensity of respiration of sporulated oocysts decreased regularly when oocysts of *E. tenella* were kept in the external environment at a temperature of 4 C for 34 days. When respiration ended completely, the oocysts probably died and lost their ability to infect. The effect of KCN on oocysts proves the importance of oxygen for sporulation. Potassium cyanide at a concentration of $10^{-3}$ inhibits sporulation, but over a long period of time this process is reversible. If respiration of oocysts is especially sensitive to cyanides which inhibit the cytochrome-oxidase system, it must mean that oocyst respiration utilizes this system.

## EFFECT OF TEMPERATURE ON OOCYST SPORULATION

Sporulation of oocysts in the external environment can occur only within a certain temperature range. The oocysts of most coccidian species can sporulate within a range of 10 to 30 C. The rate of sporogony changes when the temperature is increased or decreased. Also, the number of oocysts which develop normally depends on the temperature.

The oocysts of most species of coccidia develop well under most optimal conditions of temperature, humidity, and oxygen availability, and 85 to 95% of them usually complete sporulation. The slightest disruption of optimal conditions in the external environment changes the progress of sporulation. An oxygen deficit or an increase in temperature prevents many oocysts from sporulating or developing normally.

In most cases, at a temperature of 18 to 22 C and saturated humidity, sporulation of the oocysts is completed after a few days. In many species of coccidia the sporozoites form within 2 to 4 days after the feces has passed into the external environment. After the sporozoites are formed, another few days are required for the oocysts to become infective. This has been demonstrated with oocysts of rabbit and chicken coccidia. The oocysts of *E. intestinalis* and *E. magna,* which completed sporulation within 72 hours, did not cause an infection in rabbits, but after 120 hours the same oocysts were infective (Kheysin, 1947b, 1948). The oocysts of *E. maxima* formed sporozoites within 27 hours, but such oocysts were not infective at that time and they required another few hours to become infective (Edgar, 1954).

In each species the duration of sporulation at a certain temperature is more or less constant and is apparently determined by genotypic factors. For example, the oocysts of *E. scabra* from swine, which measure 30 by 20 microns, sporulate in 9 to 12 days. Oocysts of practically the same size (29 by 18 microns) of *E. piriformis* from the rabbit and *E. granulosa* (29 by 20 microns) from sheep sporulate under the same conditions in 36 to 48 and 72 to 96 hours, respectively. It has been observed that sometimes large oocysts sporulate somewhat slower than small ones, but this principle is not always distinctly expressed.

Large oocysts with a thick wall often require an especially long time to sporulate. The oocysts of *E. bukidnonensis,* which measured 48 by 35 microns, sporulated in 4 to 7 days at 20 C (Christensen, 1941), or in 24 to 27 days (Baker, 1938), or according to Lee (1954) in 17 days. *Eimeria leuckarti* from the horse has oocysts which measure 80 by 55 microns. At a temperature of 20 C they sporulate in 21 to 22 days. *Eimeria brasiliensis* from cattle measures 37 by 27 microns and sporulates in 6 to 14 days. The large sheep oocysts of *E. intricata* (47 by 32 microns), which have a thick wall, developed in only 3 or 4 days at 20 C, or in 1 or 2 days according to Pellérdy (1965).

One can also cite the opposite example in which the large oocysts of *E. polita* from swine, measuring 27 by 20 microns, sporulate in 8 to 9 days, while *E. intestinalis* (25 by 17 microns) and *E. media* (28 by 16 microns) sporulate in 36 to 50 hours.

Among oocysts of the same species, the small ones always sporulate faster than the large ones. For example, large *E. magna* oocysts, measuring 35 by 22 microns, completed sporulation at 22 C in 65 to 72 hours, but oocysts measuring 25–28 microns by 16–18 microns sporulated in 48 to 52 hours. Thus, oocysts of various size in one portion of feces do not all sporulate at the same time. Sporulation begins sooner in the small oocysts, and large ones of the same species may lag as much as a day behind. The same is true of other species. In rabbit feces taken at a single time, one can observe sporulation of the small oocysts of *E. media* and *E. intestinalis* within 18 hours; the oocysts of *E. magna* and the small oocysts of *E. irresidua* begin to sporulate within 24 hours, and the large oocysts of *E. irresidua* and *E. intestinalis* begin to sporulate at the end of 48 hours.

The following section lists several examples illustrating the effects of temperature on sporulation. The oocysts of *E. zurnii* normally sporulated within a range of 13 to 32 C. Some oocysts could sporulate at a temperature as low as 8 C, but the process took several months. At 10 C about 1% of the oocysts sporulated within a month; at 12 C sporulation ended between 216 and 240 hours; at 15 C, 144 hours; at 20 C, 72 hours; and at 25 C, 40 hours. Sporulation occurred most rapidly at 30 to 32.5 C and ended within 23 to 24 hours. At higher temperatures (35 to 37 C) sporulation was abnormal and somewhat suppressed. At 39 C no oocysts sporulated (Marquardt, 1957; Marquardt et al., 1960).

Similar data were obtained during studies of rabbit coccidia. Oocysts of *E. magna* and *E. perforans* did not sporulate below 10 C (Becker and Crouch, 1931). At 10 to 12 C oocysts of *E. magna* began sporulation but did not complete it in 226 to 325 hours, and 90 to 100% of the oocysts degenerated.

According to my own observations (Kheysin, 1935a), at 20 to 22 C about 95% of *E. magna* oocysts sporulated within 65 to 73 hours, but those of *E. perforans* sporulated within 48 hours. Becker and Crouch (1931) reported that at 25 C about 50% of the oocysts of *E. magna* sporulated completely within 84 hours, 40% formed sporoblasts and 10% degenerated. At this temperature, *E. perforans* required 48 hours for complete sporulation. At 33 C up to 80% of the oocysts of *E. magna* completed sporulation in 72 hours, but *E. perforans* required 48 hours. These authors reported that 33 C is the optimal temperature for sporulation of these species. At 35 C approximately 40.6% of the *E. magna* oocysts developed within 96 hours, but at 36 C sporulation began but was not completed. At the same time, 10 to 14% of *E. perforans* oocysts sporulated completely at this temperature. Oocysts of this species did not develop at higher temperatures.

Sporulation of *E. nieschulzi* took 66 hours at 20 C. At 30 C sporulation took less time, but a large number of the oocysts died. At 32 C only 39% of the oocysts developed, but at 35 C or higher sporulation did not even begin (Marquardt and Kalloor, 1962). At 20 C, oocysts of *E. tenella* sporulated within 48 hours, but at 29 C it required 15 to 18 hours for many oocysts to reach the infective stage (Edgar, 1954). Table 3 illustrates the increase in sporulation rate of chicken coccidia when the temperature was raised. Increasing the temperature to 30 C or more did not accelerate development but caused an increase in the percentage of oocysts killed. The same number of oocysts sporulated at 29 and 20 C, but at a different rate.

Edgar (1954) studied the rate of development of oocysts of *E. tenella* and their death rates at various temperatures from 8 to 40 C. At 8 C development did not occur during 9 days of observation, and after transfer to room temperature a small percentage sporulated within 4 days. At the optimal temperature of 29 C, development was completed within 18 hours, and the oocysts were infective for 7-week-old chicks. At this temperature, the pyramid stage developed within 9 hours. Within 12 hours, four sporoblasts had formed, and within 15 hours about 60% of the oocysts contained sporozoites. They became infective within 18 hours. At 28 C, the oocysts reached infectivity within 21 hours, and at 26.5 C they sporulated within 22 hours, but it took a few more hours for the oocysts to become infective. At 24 C some oocysts sporulated within 21 hours, but most completed sporulation and became infective within 30 hours. At this temperature, 70% of the oocysts developed. At 20 C 87% of the oocysts sporulated within 43 hours, and they caused infections in chicks. If oocysts of *E. tenella* were kept at a temperature of

TABLE 3
Sporulation time (in hours) of coccidia of
chickens and turkeys at two temperatures (after Edgar, 1954)

| Species | 20° | 29° |
|---|---|---|
| E. tenella | 48 | 18 |
| E. acervulina | 27 | 17 |
| E. hagani | 24–48 | 18 |
| E. mitis | 48 | 18 |
| E. maxima | 48 | 30 |
| E. necatrix | 48 | 18 |
| E. brunetti | 24–48 | 18 |
| E. adenoeides | 24 | 18 |
| E. meleagridis | 24 | 15 |

30 C, the sporozoites developed within 25 hours, and the oocysts became infective within another 2 hours. At 37 C, a small number of oocysts began to develop. Within 22 hours, 3% of the oocysts has sporulated and within 41 hours 22% had started sporulation, but they proved noninfective and sporozoites did not develop. Only after 48 hours did a small number of oocysts become infective, and by the 72nd hour most of the oocysts had died.

Sporulation of the oocysts of rabbit coccidia, as well as those of cattle and rats, is not completed at a temperature higher than 35 or 36 C. Only a small percentage of the oocysts of *E. tenella* from chickens sporulated at 37 C, but most of the oocysts ultimately died. The oocysts of *E. zurnii* began development at 37 C, but they died within 48 hours (Marquardt, 1957). This has also been observed with oocysts of *E. labbeana*. Isolated oocysts completed sporulation at 35 C, but at 37, 38, and 40 C they died without even having formed sporoblasts (Duncan, 1959a). At a higher temperature the oocysts may begin development, but it is never completed. For example, the oocysts of *E. tenella* formed sporoblasts for 3 hours at 41 C, but then development ended, and the oocysts later degenerated. At 45 C the oocysts of this species did not even begin to sporulate, and all died within 6 hours (Edgar, 1954).

The question arises as to how oocysts survive at high temperatures. There have been a series of experimental data showing the low viability of oocysts subjected to high temperatures (Becker and Crouch, 1931; Fish, 1931a; Reinhardt and Becker, 1933; Kheysin, 1937a; Chakravarty and Kar, 1946; Edgar, 1954; Duncan, 1959a; Ikeda, 1960a; Marquardt et al., 1960, and others).

Becker and Crouch (1931) reported that oocysts of *E. magna* and *E. perforans* were killed within 10 minutes at a temperature of 51 C. Partial death of oocysts was observed at 50 C with the same amount of exposure. Fish (1931a) studied the effects of high temperatures on the oocysts of *E. tenella*. It took the following lengths of time to bring about death of all of the oocysts:

| Temperature in C | Time |
|---|---|
| 45 | 24 hr |
| 50 | 1.5 hr |
| 53 | 10 min |
| 55 | 3 min |
| 60 | 15 sec |
| 70 | 15 sec |
| 80 | 5 sec |
| 90 | 5 sec |

Thus, a 10 minute exposure at 53 C killed the oocysts of *E. tenella*. The unsporulated oocysts of *E. miyairii* from rats were killed in 10 minutes at 49 C (Reinhardt and Becker, 1933). In some experiments oocysts died at this temperature in as little as 7 minutes. At a temperature of 50 C it took only a 4-minute exposure to kill all the oocysts of *E. miyairii*. At 51 C all oocysts died within 2 minutes, at 52 C, in 1 minute, and at 53 C , in 15 seconds.

According to Chang (1937) the oocysts of three species of rat coccidia, as well as *E. tenella* from chickens, differed markedly from one another in their sensitivity to high temperatures (Table 4). It is apparent from this table that oocysts of *E. separata* were the most sensitive, followed by *E. nieschulzi* and *E. miyairii*. The oocysts of *E. tenella* were more resistant to high temperatures than the oocysts of all the species of rat coccidia. Chang discovered that of sheep coccidia, the most resistant to high temperatures was *E. arloingi*, followed by *E. faurei* and *E. ninakohlyakimovae* (Table 5). From Table 5 it is apparent that sheep coccidia were generally less sensitive to high temperatures than oocysts from the rat and *E. tenella* oocysts. The most resistant was *E. arloingi*, and the most sensitive was *E. separata*. *Eimeria tenella* was less sensitive than *E. miyairii* and *E. nieschulzi*. A 100% mortality after 10 minutes of

### TABLE 4
Time of death (in minutes) of 50 per cent of oocysts of rat coccidia and
*E. tenella* from chickens (after Chang, 1937)

| Temperature in ° C | E. miyairii | E. nieschulzi | E. separata | E. tenella |
|---|---|---|---|---|
| 56 | 0.41 | 0.016 | 0.01 | 0.11 |
| 54 | 1.4 | 0.09 | 0.13 | 0.30 |
| 52 | 6.1 | 0.27 | 0.26 | 2 |
| 50 | 17.5 | 1.7 | 0.36 | 16 |
| 48 | 58 | 12.2 | 0.78 | 83 |
| 46 | 138 | 42.5 | 13.8 | 245 |

### TABLE 5
Time of death in minutes of 50 per cent of oocysts of sheep coccidia

| Temperature in ° C | E. arloingi | E. faurei | E. ninakohlvakimovae |
|---|---|---|---|
| 56 | 0.51 | 0.42 | 0.32 |
| 54 | 2.1 | 1.6 | 0.78 |
| 52 | 14.4 | 7.9 | 7.1 |
| 50 | 29 | 19.8 | 9.2 |
| 48 | 180 | 130 | 81 |
| 46 | 540 | 542 | — |

exposure was also different in *E. miyairii* and *E. tenella*, and in the former species it occurred at 49 C, and in the latter at 53 C. In *E. magna* from rabbits the critical point was 51 C.

It is possible that such a difference in sensitivity to high temperatures is determined by the varying thicknesses of the walls. In *E. miyairii*, the wall is more delicate than in the other two species. It does not appear possible to explain the difference on the basis of oocyst size. It is possible that these differences depend on the varying resistance of the proteins in the cytoplasm of the zygote to damaging external agents. However, there is, as yet, no direct evidence for this hypothesis.

According to Chang (1937) sensitivity of oocysts to high temperatures was, to a considerable degree, associated with the diet on which the experimental animals were kept. Oocysts of *E. nieschulzi,* isolated from rats receiving a diet deficient in vitamins B and C for 9 days prior to infection were more sensitive to high temperatures than those isolated from rats on a normal diet (Table 6). If the rats were given a surplus of

TABLE 6

Time of death (in minutes) of 50 per cent of the oocysts of *E. nieschulzi* from rats on different diets (after Chang, 1937)

| Temperature in ° C | Avitaminosis | Control |
|---|---|---|
| 52 | 0.18 | 0.32 |
| 50 | 0.73 | 1.4 |
| 48 | 2.4 | 15 |
| 46 | 12.5 | 78 |

vitamins B and C, this did not alter the sensitivity of the oocysts to high temperatures. Starvation for 3 days after infection likewise caused no change in sensitivity to high temperatures. These data, to a certain degree, indicate that heat resistance in oocysts more likely depends on the status of the protoplasm in the zygote rather than on the character of the oocyst wall.

The development of oocysts and their viability at low temperatures has been studied by few investigators. Among the rabbit coccidia, only the oocysts of *E. perforans* sporulated at 10 C (Becker and Crouch, 1931). Other species did not develop at this temperature. At a temperature of 5 C, unsporulated oocysts of *E. stiedae* survived 3.5 to 4.5 years in a physiological solution, after which they sporulated at room temperature in a 2% solution of $K_2Cr_2O_7$ (Hagan, 1958).

Oocysts of *E. labbeana* from pigeons were stored for 60 days at 2 to

3 C and sporulated under normal conditions after a slight time lag. However, after 60 days at a temperature of 0 to −2.5 C, no oocysts remained alive; but 30 days in the same environment did not kill them (Duncan, 1959). However, Chakravarty and Kar (1946) reported that oocysts of this same species did not sporulate at room temperature after 168 hours at 4 C.

Edgar (1954) placed unsporulated oocysts of *E. tenella* in an environment of −12 C for various lengths of time. After a 6-hour exposure the oocysts sporulated normally. However, 7 days at this temperature caused the death of all the oocysts.

Sporulated oocysts of *E. tenella* were subjected to the effects of a temperature from −2 to −5 C for 7 days. After this time the oocysts were given to chicks; however, no infections followed. A 4-day exposure at the same temperature did not kill the oocysts, and they caused a severe infection (Uricchio, 1953b). The oocysts of *E. stiedae* died quickly at a temperature of −10 and −15 C (Pérard, 1925; Hagan, 1958).

Oocysts of swine coccidia are somewhat more resistant to low temperatures than oocysts of *E. stiedae* and *E. tenella*. The oocysts of *E. debliecki* and *E. scabra* did not sporulate within 26 days at a temperature of 6 to 8 C, but within 14 days after being returned to room temperature (19–25 C) 88% of them completed sporulation. In other experiments, oocysts in water were subjected to temperatures from 0.5 to −3 C and from −2 to −7 C for 26 days. Again development did not take place, but the oocysts retained their viability, and within 14 days after being returned to room temperature, 67 and 55%, respectively, completed sporulation (Avery, 1942).

Oocysts of *E. zurnii* remained undamaged after being frozen in water at a temperature of −7 to −8 C for 2 months. Within 2 weeks after being returned to room temperature, 92% of the oocysts sporulated after being placed in optimal surroundings. After an exposure lasting from 4 weeks to 2 months, 74% of the oocysts sporulated. In another experiment at the same temperature, 50% of the oocysts survived a 5-month exposure. At −30 C, only 36% of the oocysts survived for 6 hours and 5% lasted 24 hours. If the oocysts were desiccated on glass and then placed in a temperature of −30 C, 64% survived for 6 hours, and 8% for 24 hours (Marquardt et al., 1960). At −20 C for 2 days, about 50% of the oocysts of *E. zurnii* survived (Petrov and Nikonov, 1964). A small percentage of the oocysts of *E. bovis* remained viable at a temperature of −5 to −7 C after about a 1-month exposure (Wilson and Morley, 1933).

Christensen (1939) reported that in pure water at a temperature from

0 to 5 C, oocysts of *E. arloingi* developed in 2 weeks, but at room temperature it took 2 or 3 days for sporulation. At low temperatures and in a suspension of feces, approximately 20% of the oocysts were viable for 19 months, but they did not sporulate.

Landers (1953) conducted detailed studies on the influence of low temperatures on sheep coccidia. He froze sheep feces containing large numbers of oocysts. At a temperature of $-40$ and $-70$ C, the oocysts died very rapidly. A 2-hour exposure caused 100% mortality. There were no positive effects on the viability of the oocysts if they were first kept at $-19$ C for 24 hours and then placed in a freezer at $-40$ or $-70$ C; all died within 2 hours. Freezing at $-30$ C for 4 hours caused the death of only 50% of the oocysts, and a 24-hour exposure killed them all. However, if the oocysts were first kept for 24 hours at $-19$ C and then placed in a freezer at $-30$ C for 24 hours, 95% remained viable. However, a 48-hour exposure was fatal to all oocysts. After storing at $-25$ C for 1 to 7 days after preliminary cooling to $-19$ C or at $-25$ C without prior cooling, 90 to 95% of the oocysts remained viable. A 14-day exposure killed 50% of the oocysts of *E. arloingi* and *E. ninakohlyakimovae*. The oocysts of *E. parva* proved more resistant. Approximately 80% of them died after being frozen. A temperature of $-19$ C for 60 days did not cause the oocysts to die. Multiple freezing and thawing at $-19$ C did not cause the death of the oocysts. After being frozen seven times in 4 days, 95 to 98% of the oocysts remained viable. The same was observed after five freezings of sheep coccidia oocysts down to $-25$ C for 8 days; 83 to 94% of the oocysts remained viable. Freezing six times over a period of 9 days did affect the viability of the oocysts; 35% of *E. arloingi*, 50% of *E. ninakohlyakimovae*, and 90% of *E. parva* died.

Thus, for the oocysts of three species of sheep coccidia, a temperature of $-25$ C was near the minimum at which they could survive. A temperature of $-30$ C or lower was lethal even after relatively short exposures. Landers (1953) reported that such resistance to low temperatures makes it easy to understand the cause of high viability of oocysts which lie for long periods of time on the surface of the soil of pastures during the wintertime.

## RELATIONSHIP BETWEEN SPORULATION AND HUMIDITY

Coccidian oocysts from terrestrial hosts are usually passed with the feces onto the soil. One of the basic conditions needed for normal sporulation of oocysts is sufficient moisture in the environment. Oocysts develop best under conditions of saturated humidity and free access to oxygen. A humidity deficit causes a wrinkling of the oocyst wall due to water loss. As a result, the zygote is pressed by the collapsed walls, and sporogony cannot proceed normally.

The oocysts of rabbit coccidia survived well after a loss of moisture from the feces. They remained viable for 10 days or more in dry feces. Dehydrated oocysts of *E. magna* were severely deformed after 5 days at room temperature. However, after transfer to distilled water, they re-gained their normal shape and within 12 to 20 days, 30% of the oocysts sporulated and became infective (Kheysin, 1935a). The oocysts of *I. felis* and *I. rivolta* tolerated dehydration of the feces for 5 days at 16 C, and 10% of the oocysts sporulated normally. Dehydration at 36 C caused the death of the oocysts, but 50% of them were still able to form sporoblasts (Kheysin, 1937a).

The oocysts of *E. zurnii* developed at 25% relative humidity and sporulated in 3 days at 30 C. However, only 12% of the oocysts formed sporozoites; the remainder died. At 50% humidity, as high as 30% of the oocysts sporulated, and at 75% humidity, 27 to 51% sporulated. At temperatures lower than 16 C and a relative humidity of 25%, about 9% of the oocysts sporulated in 12 days. At 75% relative humidity, 21 to 56% of the oocysts developed (Marquardt, 1960b).

The death of *E. tenella* oocysts from a moisture deficit depended upon the temperature. The higher the temperature, the greater the number of oocysts which died from a decrease in relative humidity (Ellis, 1938). When oocysts of *E. labbeana* were dried at room temperature for 3 days, they retained their ability to cause infections, but after a 4-day exposure all died (Duncan, 1959a). The relatively high resistance of oocysts to a moisture deficit can be explained by the structure of the protein wall through which water cannot easily penetrate. However, dehydration was one of the most important lethal factors to oocysts in the external environment (Pérard, 1925; Wilson and Morley, 1933; Ellis, 1938; Christensen, 1939; Brotherston, 1948; Farr and Wehr, 1949; Duncan, 1959a). Oocysts of *E. tenella* survived a week in dried feces, but drying in an incubator at 25 C killed them in 1 hour (Pellérdy, 1965).

It is interesting that the oocysts of coccidia from animals living in water, such as *E. carpelli* from the carp, are wholly unadapted to dehydration. It takes only 1 hour outside of water to kill all oocysts (Zaika and Kheysin, 1959).

It should be pointed out that deformation of the oocysts does not necessarily cause the death of the oocysts. At a humidity of 45%, oocysts of *E. labbeana* survived at a temperature of 20 C for 3 days, even though externally they appeared completely deformed. Under such conditions, the viability of the oocysts should be determined by introducing them into animals to produce an infection (Duncan, 1959a and b).

## RELATIONSHIP BETWEEN SPORULATION AND THE EFFECTS OF CHEMICAL FACTORS

Oocysts in the external environment may be under conditions where they come under the influence of various chemical substances which, to one degree or another, may penetrate into the oocyst and influence the zygote. The penetration of chemical substances into the oocyst is determined by the properties of the oocyst wall. Because of the way in which the walls are constructed, many chemical substances cannot penetrate inside the oocyst and thus do not impede sporulation. The internal protein wall does not dissolve in either strong acids or bases, and being semipermeable, it apparently accounts for the protection of the zygote against the passage of various chemicals. With the use of oocysts of rabbit coccidia from which the external wall had been artificially removed, it was determined there was no difference in resistance to chemical substances as compared with intact oocysts (Kheysin, 1935a and b). This demonstrated the primary role of the internal wall in the protection of the zygote and sporozoites against the effects of various chemical substances. Apparently, the external wall protects only against mechanical damage and does not hinder the entry of chemical material into the oocyst.

In most cases, the action of chemical substances has been determined by use of unsporulated oocysts, and the criterion for penetration of substances into the oocysts was their ability to sporulate. However, in some cases, the viability of the oocysts was determined by their ability to infect susceptible animals (Duncan, 1959a).

Often when under the influence of some chemical substances, the oocysts are severely deformed, as happens when they lose moisture. Such oocysts may sometimes be mistakenly classed as dead. It is highly probable that this is the reason for disagreements in the evaluation of results obtained. For example, Bondois (1936) noted that the oocysts of *E. pfeifferi* ( = *E. labbeana*) died in a 15% solution of NaCl in 24 hours, but according to Duncan (1959a) the oocysts of this species survived for 10 days in the same solution. This disagreement resulted from the fact that Bondois based his results only on the external appearance of the oocysts without checking their actual ability to cause an infection in animals. The second author determined the true state of the oocysts by using infectivity as a criterion. The sporulated oocysts had become wrinkled from water loss after 10 days in the 15% salt solution, but they succeeded in infecting pigeons (Duncan, 1959a). Thus, errors in methodology may serve as a source of serious errors in the study of the effects of various chemicals on sporulation and viability.

In evaluating the effects of a given preparation, it should also be kept in mind that even under the best developmental circumstances, such as a 2% potassium bichromate solution, not all the oocysts will sporulate. Approximately 90 to 95% of the oocysts sporulated normally if the experimental animal receiving the oocysts was not immune to a given species of coccidia (Kheysin, 1935a, b, and c, 1947b; Duncan, 1959a).

Plasmolysis of the oocysts of *E. magna, E. intestinalis,* and other species has been observed in hypertonic solutions of salts (Kheysin, 1935a), but the oocysts were not able to sporulate until there was a complete loss of water from the zygote and the oocyst. In weak concentrations of salts, such as 1 N NaCl, KCl, LiCl, $CaCl_2$, $MgCl_2$, $BaCl_2$, $KNO_3$, $Na_2SO_4$, $AgNO_3$, $ZnSO_4$, $K_4Fe(CN)_6$, $CoCl_2$, $CuSO_4$ and others, plasmolysis began in 4 or 5 days, but the oocysts of rabbit, cattle *(E. zurnii),* sheep *(E. arloingi),* and cat *(I. felis* and *I rivolta)* coccidia sporulated normally. In 5 N or in saturated salt solutions, plasmolysis began in 24 hours in oocysts of *E. magna* and others. Within 5 days the oocyst wall was bent so that the oocyst looked like a little boat with strongly warped gunwales. Transfer of such oocysts to water led to rehydration in 15 days. When this was done, 90% of the oocysts sporulated. The plasmolysis which took place showed that the salts did not penetrate into the oocyst.

Oocysts of *E. bovis* sporulated in a $10^{-4}$M solution of $Na_2S$, $10^{-2}$M $Na_2SO_3$, $NaNO_3$, and 0.1 M $NH_4OH$ (Senger, 1959). In various experiments approximately 20 to 80% of the oocysts completed sporulation in these solutions. Pérard (1924) reported that oocysts of *E. stiedae* and *E. magna* did not sporulate in a 0.1% $KMgO_4$ solution. Oocysts of *E. tenella* did not develop in a 5% KI environment (Horton-Smith et al., 1947). According to Kar (1947), the oocysts of *E. stiedae* and *E. perforans* survived for 24 hours in a 10% solution of $CuSO_4$.

Solutions of $Na_2CO_3$ and $NaHCO_3$ had a special effect on oocysts. Normal solutions of these caused a highly irregular sporulation of oocysts of *E. magna, E. irresidua,* and *E. perforans* after 3 to 5 days, or stopped them from sporulating altogether. Abnormal sporulation resulted in the formation of two sporoblasts instead of the usual four and the appearance of large vacuoles in the zygote. After 3 days in a solution of 1 N $Na_2CO_3$, 50% of the oocysts formed only oval sporoblasts.

The oocysts of *E. labbeana* sporulated very slowly in 2% $NaHCO_3$, and in a 5% solution they did not sporulate at all. However, after 10 days in a 2% $NaHCO_3$ solution, they were still able to cause infections (Duncan, 1959a and b).

Carbon dioxide and ammonia easily penetrate oocysts and cause a cessation of development. In an atmosphere of carbon dioxide the oocysts do not develop, but degenerate and undergo irregularities in development. If unsporulated oocysts of *E. magna* were placed in a 25% solution of ammonia, a 30-minute exposure was sufficient to stop sporulation. This process was not reversible. Weak solutions of ammonia (a 10% solution) slowly penetrated oocysts and did not retard development. Petrov and Nikonov (1964) reported that after 10 minutes of exposure to ammonia, oocysts of *E. zurnii* lost their ability to sporulate. A 1% solution of ammonia killed 100% of the oocysts of *E. tenella* in 24 hours, a 5% solution was lethal in 2 hours and a 10% solution in 45 minutes (Horton-Smith et al., 1947).

It is highly probable that the high death rate of *E. tenella* oocysts (and other species) in the external environment, such as in litter on cattle paths, is determined not so much by a moisture deficit as by the action of ammonia, which accumulates in the litter when manure is broken down (Horton-Smith, 1947, 1954). Because of this, some investigators recommended using a method of fumigating the litter with ammonia to destroy oocysts (Horton-Smith et al., 1940). This same method is recommended for fumigating buildings in which chicks are kept. It has been shown experimentally that ammonia at a concentration of 25.0 mg/m$^2$ destroys 100% of the oocysts in 1 hour, and a concentration of 7.0 mg/m$^2$ destroys oocysts in feces within 3 hours. From a practical standpoint this application is possible for destruction of oocysts in closed buildings.

The mercury salts, mercuric cyanide, and mercuric chloride, penetrate through the oocyst wall and inhibit development. In 35 minutes, a saturated solution of mercuric chloride killed oocysts of *E. stiedae, E. magna,* and *E. perforans.* Oocysts of the last species began development after being placed in $K_2Cr_2O_7$, but sporozoites did not form. A 1-hour exposure of *E. perforans* to mercuric chloride killed all oocysts. In solutions of 0.2, 0.1, and 0.01 N $HgCl_2$, development was retarded as compared with controls, and only isolated oocysts formed sporozoites. In a solution of 1:1000 HgCl$_2$ for 72 hours, only 40% of the oocysts of *E. magna* and *E. perforans* completely sporulated. In 0.02 and 0.001 N solutions, oocysts of these species developed in 96 hours, but 65% of the oocysts died in various stages of sporulation (Kheysin, 1935a, b, and c). Using oocysts of *E. tenella,* Fish (1931a and b) obtained similar data. He found that a 1% solution of HgCl$_2$ killed all of the oocysts in 48 hours, but a 0.1% solution caused death of 18.4% of the oocysts in the same period of time.

Oocysts of *E. zurnii* from cattle sporulated in a $10^{-6}$M solution of mercuric chloride, but sporozoites formed in only 3% of the oocysts (Marquardt et al., 1960). Approximately 80% of *E. bovis* oocysts died in a $10^{-6}$or $10^{-8}$solution of HgCl$_2$ (Senger, 1959). According to Yakimov and Galuzo (1927), the oocysts of cattle coccidia did not die in a 3.7 × $10^{-4}$ M concentration of mercuric chloride. Mercuric cyanide, like mercuric chloride, penetrated the oocysts of *E. magna* and impeded their development.

The toxicity of mercuric chloride can be lowered by the addition of certain salts such as KCl, KI, NaCl, and CaCl$_2$. The addition of an equal volume of a saturated solution of one of the above salts to a saturated solution of mercuric chloride allowed all oocysts to develop normally (Kheysin, 1935a). If the proportion of salts was decreased as compared with the amount of mercuric chloride, the percentage of dead oocysts increased. Some salts, such as Na$_2$SO$_4$, KNO$_3$, NaNO$_3$, and BaCl$_2$, did not reduce the toxicity of mercuric chloride. An analogous decrease in the toxicity of mercuric chloride was observed long ago by certain microbiologists. They noted that a mixture of HgCl$_2$ and NaCl decreased the toxicity of mercuric chloride for spores of anthrax bacteria.

It is highly probable that the mechanism of this phenomenon is that mercuric chloride and some salts form compounds which have a different physiological activity and do not penetrate through the wall of the oocyst. For example, the addition of KI to mercuric chloride leads to the following reaction: $4KI + HgCl_2 = 2KCl + K_2HgI_4$. The last complex salt forms a K ion and the HgI ion complex. Such ions do not penetrate the oocyst. If a small amount of salt is added, the mercuric ions remain in the solution and prevent part of the oocysts from developing. Na$_2$SO$_4$ joins with HgCl$_2$ in an exchange reaction and free mercury ions remain in the solution and impede the development of the oocysts (Kheysin, 1935a and b).

Sporulation of oocysts of *E. bovis* was blocked by the addition of potassium cyanide and sodium azide, i.e., by substances which inhibit the cytochrome system of the cell. Apparently, such a system exists in oocysts, as a result of which, substances blocking the cytochrome system stop the development of oocysts. Sporulation was also retarded by the presence of 2,4-dinitrophenol, which inhibited oxidative phosphorylation (Senger, 1959).

The inorganic acids HCl, $HNO_3$, and $H_2SO_4$ do not penetrate oocysts in any concentration incapable of breaking the wall. The sporulation time of *E. magna* is not decreased in the presence of 1 N solutions of these acids. Galli-Valerio (1918) reported that *E. zurnii* oocysts did not sporulate in 5% HCl. Yakimov and Galuzo (1927) found that oocysts of this species developed in 2% $H_2SO_4$, but in a 3 to 5% solution of sulfuric acid they formed only sporocysts and not sporozoites. Pérard (1924) noted that 20% sulfuric acid stopped the sporulation of *E. magna*, but according to Kheysin (1935a) such a concentration had no effect on the normal course of sporulation in this species of rabbit coccidia. Oocysts of *E. labbeana* sporulated with some lag in 2 to 5% HCl (Duncan, 1959a). Only 5% of the oocysts of *E. tenella* died in 2 N HCl (Fish, 1931a).

Bases do not penetrate the undamaged oocyst wall. A 2 N and 5 N solution of KOH and NaOH caused a strong, reversible plasmolysis in *E. magna* in 72 to 120 hours but did not hinder sporulation. Twenty days in 5 N KOH did not have a lethal effect on oocysts of rabbit coccidia. Yakimov and Galuzo (1927), Galli-Valerio (1918), and Pérard (1925) found that the oocysts of various species *(E. magna, E. stiedae,* and *E. zurnii)* died quickly in a 10% solution of KOH. According to Pérard, *E. perforans* sporulated in 10% KOH, but not in a 20% solution. About 10% of *E. stiedae* oocysts died in a 2% solution of NaOH after 24 hours of exposure. Oocysts of *E. tenella* developed in 0.5 to 2 N solutions of NaOH in 48 hours, and the death rate in these solutions did not exceed 1.1% (Fish, 1931a). Oocysts of *E. tenella* did not develop in a 5% solution of KOH. Oocysts of *E. bovis* developed in 2.5 M solutions of NaOH and sulfuric acid (Senger, 1959).

Organic substances effect oocysts in various ways. In large concentrations, sugar does not retard sporulation. Retardation may occur only if the sugar undergoes serious putrefaction resulting in an oxygen deficit. Solutions of urea, 10 and 20%, retarded the onset of sporulation and killed about 30% of the oocysts of *E. magna*. When oocysts were sporulated in rabbit urine, they matured somewhat later than controls. The same was noted in digestive juice and in bile at a temperature of 22 C.

Strychnine, 1% nitric acid, hydrochloric acid, morphine, and cocaine do not retard sporulation of oocysts. Low concentrations (normal solutions) of organic acids, e.g., lactic, succinic, and oxalic acids, do not hinder sporulation. At higher concentrations (up to 5 N) these acids cause plasmolysis in 72 hours without retarding sporulation of the oocysts of *E. magna, E. perforans,* and *E. stiedae.* After a 72-hour exposure, sporulation was not completed by any oocysts of *E. magna* or *E. perforans* in 5 M formic acid. After a 3-hour exposure 85% of the oocysts of *E. magna* sporulated; after a 24-hour exposure only 50% sporulated, and after 48 hours, less than 26% sporulated. In 5 M acetic acid 30% of the oocysts sporulated after 72 hours; 10% sporulated in 124 hours, but all oocysts of *E. magna* and *E. stiedae* died after 124 hours in formic acid. Thus, it appears that formic acid penetrated the oocyst wall faster than acetic acid.

A 40% solution of formalin killed the oocysts of rabbit coccidia in 24 hours, but a 4% solution caused no damaging effects (Kheysin, 1935a). Pérard (1925) found that oocysts of rabbit coccidia developed in 5% formalin. In a 1% solution, the oocysts of *E. magna* retained their viability for several months (Kheysin, 1935a). Orlov and Aytykina (1936) reported that oocysts of rabbit coccidia sporulated in a 1 to 10% solution of formalin when clumps of feces containing the oocysts were added to the solution. Fish (1931a) observed that after 48 hours of exposure to 2 to 5% formalin, approximately 60 to 70% of the oocysts of *E. tenella* completed their development normally. Senger reported the same for *E. bovis.* Oocysts of *E. zurnii* did not sporulate in a 0.25 M solution of formalin, and in a 0.01 M solution only 8% of the oocysts completed development. According to some data, the oocysts of *E. labbeana* can sporulate in 2.5% formalin, but Duncan (1953) did not observe sporulation of oocysts of this species in 2 or 5% formalin; however, they sporulated normally in a 1% solution.

Alcohol inhibits sporulation to a considerable degree. Oocysts of *E. magna* suffered plasmolysis after approximately 8 to 10 hours in ethyl alcohol. After transfer of such oocysts to water, rehydration and sporulation began. In absolute methyl alcohol, 100% of *E. magna* oocysts died in 10 to 14 hours. In absolute ethyl alcohol, the same species all died in 18 to 24 hours. In *n*-propyl alcohol the oocysts died within 36 hours, but in isobutyl alcohol this took 96 hours. In many oocysts development began, but fragmentation of the sporoblasts developed and death followed.

Of the equimolecular solutions (such as 15 M [sic]), methyl alcohol is the quickest to cause retarded sporulation and death. Within 24 to 48 hours, 48% methyl alcohol killed all *E. magna* oocysts. A 69% solution of ethyl alcohol had the same effect in 48 to 72 hours, as did 90% *n*-

propyl alcohol. In absolute isobutyl alcohol, death occurred in 96 hours, but some oocysts survived this treatment and formed sporocysts (Kheysin, 1935a). In weak solutions (20 to 30%) of ethyl alcohol, oocysts of *E. magna* formed sporozoites, but 50% of them plasmolyzed within 48 hours, and development did not continue. Glycerol penetrated the oocysts of *E. magna* slowly, and at first caused detachment of the walls, then plasmolysis. After 72 hours in glycerol, sporulation was not retarded.

Chloroform, ether, acetone, and benzol penetrate oocysts rather quickly and impede sporulation. Oocysts of *E. magna* died in chloroform within 2 or 3 hours; in ether they died in 5 to 6 hours; in acetone, in 8 hours; and in benzol, in 8 to 10 hours. No oocysts underwent sporulation after such treatment.

Various cresol compounds, phenol, and xylene penetrate oocysts in 6 or 8 hours and cause complete disruption of sporulation. Lysol and phenol concentrate between the external and internal walls of the oocyst and gradually diffuse inside.

Pérard (1924) reported that 2% carbolic acid slowed the development of *E. magna* oocysts in 24 hours. A 5% carbolic acid solution killed the oocysts of *E. tenella* in 20 to 48 hours (Tyzzer et al., 1932). According to Fish (1931a), 2 or 5% solutions of phenol or cresol caused the death of oocysts of *E. tenella* in 48 hours. Sporulation of *E. zurnii* was completely suppressed by 0.05 M phenol. In a solution of 0.01 M phenol, only 8% of the oocysts developed (Marquardt et al., 1960). A 1% emulsion of creolin did not retard sporulation of rabbit coccidia oocysts, but 2.5% killed approximately 50% of the oocysts.

The oocyst wall is impermeable to many staining agents, some of which are used for vital staining of cells. The oocysts of *E. magna* sporulate normally in any concentration of chrysoidine, neutral red, methyl blue, methyl green, eosin, orange, fuchsin, or trypan blue, and the zygote does not stain. The same is observed in oocysts killed by ammonia.

Thus, sporulation of the oocysts may be completed in solutions of many substances. This is because the oocyst wall is semipermeable and consists of very dense proteins. The substances which penetrate the oocyst most rapidly are those which dissolve lipids or are dissolved in lipids. It would seem that the presence of lipids in the wall also determines their high resistance to penetration. However, as was mentioned earlier, it is not always possible to demonstrate the lipids which make up the wall by cytochemical methods. It is highly probable that certain substances penetrate the oocyst mainly because of their molecular size, and not because of their solubility in lipids. Actually, the substance which penetrates the wall most rapidly is methyl alcohol, followed by ethyl, *n*-propyl, and isobutyl alcohols. Each member of this series has a greater molecular volume than the previous one. This type of penetration into the oocysts is undoubtedly determined by the physical properties of the internal protein wall (Kheysin, 1935a).

Data on the effects of various substances on oocysts have a definite significance for purely practical purposes. The effectiveness of prophylaxis in controlling coccidiosis of animals may be affected considerably by the influence of chemical compounds on oocysts in the external environment. It is apparent that it is very difficult to slow down sporulation. Very few chemical substances lethal to oocysts can be used for disinfection in stalls, bird roosts, and on pastures. Because they have little effect on them, a large number of the common disinfectants are not very suitable for destroying oocysts. One of the important and immediate goals should be the search for new means of quickly penetrating the oocyst and halting the process of sporulation. Kar (1947) reported that rabbit coccidia oocysts can be destroyed after a 24-hour exposure to a 9% solution of monochloroxylenol or a 3% solution of terpineol. Ohkubo, et al. (1955) stated that a 1.5% emulsion of *o, o*'-dichlorbenzene used for 5 hours at room temperature will kill 100% of the oocysts of various chicken coccidia.

## SPORULATION OF THE OOCYSTS IN THE EXTERNAL ENVIRONMENT

The oocysts and feces enter the external environment where actual sporulation occurs. The terrestrial animals leave their feces on the surface of the soil, and the coccidia of aquatic organisms naturally end up in water where their existence is determined by the conditions of the water. Sporulation and viability of coccidian oocysts of land animals also depends on conditions which occur at the soil surface in various biotopes and in various geographic zones. Sporulation of oocysts in the external environment depends on three basic factors: temperature, humidity, and free access to oxygen. The possibility of oocysts development on the soil and their ultimate viability will not be identical in all geographic zones, on all soils, in different types of illumination, or at all times of the year. This has been shown in a series of investigations conducted in different countries (Patterson, 1933; Horton-Smith, 1947; Farr and Wehr, 1949; Kogan, 1956, 1959; Krylov, 1959a, 1960).

Tyzzer et al. (1932) reported the rapid death of chicken coccidian oocysts in the soil. However, Patterson (1933) succeeded in infecting chicks with oocysts which had been preserved in the soil for 21 weeks Delaplane and Stuart (1935) observed live oocysts of *E. tenella* in the soil of shaded meadow plots 1.5 years after removal of chickens. According to Boughton (1939), the survivability of chicken coccidia on the surface of the soil or at a depth of 5 cm depended on the degree of moisture and the intensity of putrefaction. If the feces dried out, no more than 20% of the oocysts sporulated in 2 to 6 days.

Levinson and Fedorov (1936) found that the oocysts of rabbit coccidia matured on moist sand but not in loam. The authors explained this difference as being due to insufficient aeration in loam. Horton-Smith (1954) and other authors have shown that the oocysts of chicken coccidia on the soil and in litter died chiefly from the effects of ammonia which is formed when manure decomposes. A great deal of ammonia is formed deep in the litter where the oocysts died in 24 hours, even under optimal conditions of temperature and humidity.

Detailed studies on the rate of sporulation and survivability of oocysts of *E. tenella* and *E. necatrix* on the soil in the Mogilev Oblast' of the Belorussian SSR were conducted by Kogan (1956, 1959). The experiments were conducted from May through July. Unsporulated oocysts of *E. tenella* and *E. necatrix* placed in sand, clay, and soil at a depth of 4 cm formed sporozoites in 1 week when they received water every day. However, some development was observed in 54 to 61% of the oocysts. Within 2 weeks, this percentage had increased to 75 to 84%. Oocysts buried 9 cm in the soil also sporulated in 1 week. Thus, in Belorussia oocysts of chicken coccidia develop normally in the summertime.

In the same locality, oocysts of *E. tenella* and *E. necatrix* in feces were placed in the soil at a depth of several centimeters in November and left under the snow all winter. Oocysts left on the surface of the soil where there was no snow cover from November through April did not sporulate even after being transferred to a 2% $K_2Cr_2O_7$ solution and only 1.7% of the oocysts formed sporocysts. The same results were obtained when the oocysts were kept under 5 to 30 cm of snow on the surface of the soil. After the oocysts were transferred to optimal surroundings, only 1.1% developed.

In the same period, from November through March, and under the same conditions, sporulated oocysts were found. In spring they were used to infect previously uninfected chicks. All infections were of a low intensity and produced few oocysts. Accordingly, it may be said that 5 months of winter kills many oocysts, but some of them remain viable and retain their ability to invade. On the pastures and poultry farms of Belorussia, infective oocysts were present in the spring, creating great possibilities that chicks could become infected.

In the Moscow Oblast' unsporulated oocysts of *E. tenella* survived on the surface of the soil of pastures for 7 months from autumn to spring (Machinsky, 1954). In the study of sporulation of *E. arloingi* oocysts and other species of sheep coccidia in the high-mountain valleys and slope pastures of Tadzhikistan, different results were obtained by Krylov (1959a, 1960). On the surface of soil not shaded by brush in the high-mountain pastures, the oocysts did not sporulate in 22 days from January through February. The temperature of the soil during this period averaged $-2.3$ C but on some days it warmed to 11.5 C. When higher temperatures occurred in January, as was observed in 1958, 6% of *E. arloingi* oocysts sporulated in 12 days. Christensen (1939) reported the possibility that oocysts of this species may develop at a temperature of about 5 C.

In May, the soil temperatures on the high mountain pastures of Tadzhikistan are so high that the humidity drops considerably. However, 4% of the oocysts on the surface of the soil succeeded in sporulating in 2 days, but they then died from the effects of the high temperatures, which reached 56 C (a lethal level for oocysts). In July and August, despite the high temperature on the surface of the high mountain pastures (up to 45 C), about 30% of the oocysts sporulated on the 5th day if they were under grass cover. Because of the high humidity during this period, the destructive effects of high temperatures leveled off. The mountain meadow soil, with its patches of abundant vegetation, is capable of retaining moisture for long periods of time. But even under these conditions most of the sporulated oocysts died in 8 to 10 days (Krylov, 1959a). Under natural conditions, the oocysts died mainly because of a lack of moisture or high temperatures.

Shevchenko (1953) noted that on the steppe pastures of the Chkalov Oblast', the oocysts of sheep coccidia died in 8 hours when the soil temperature ranged between 26 and 34 C. Orlov (1956) thought that during the winter, the oocysts of sheep coccidia on the open steppe pastures of Kazakhstan died because of partial freezing and thawing. Christensen (1939) observed survival of sheep coccidia in the soil during the cold part of the year in the USA. Melikyan (1954) stated that in Armenia, at places where sheep were kept on milking plots and on pastures, the oocysts of sheep coccidia survived for more than a year.

Krylov's work in Tadzhikistan has shown that oocysts of *E. arloingi, E. faurei, E. intricata,* and others retain their viability for several winter months. Oocysts left on low grass, semi-savanna pastures in November were still viable until January of the following year, and about 70% of the oocysts sporulated under such conditions. The winter temperature averaged 11 to 14 C. In colder winters more oocysts died. Frequent thawing and freezing of feces containing oocysts was especially harmful. Levinson and Fedorov (1936) reported a rapid death of rabbit coccidia in feces buried in water at low temperatures.

In the spring, grass appears on the low grass pastures of Tadzhikistan, and the temperatures of the soil surface reaches 31 to 35 C. Because of the lack of rain and the high temperature of the surrounding environment during this period, the oocysts die faster than in winter. Almost all oocysts dropped in winter under these conditions died by April. In May, all the oocysts die because the temperature rises to 60 C and the feces lying on the soil under the grass dries out completely. In summer, the oocysts on the soil die even faster than in winter and spring, because of the high temperature and low humidity. If oocysts get into the soil in May, they are completely destroyed by June because of the lack of water. During the summer the pasture is practically free of oocysts.

During the summer no such high temperatures are observed in the high mountain, alpine pastures as occur on the low ones. But despite sufficient humidity, the oocysts dropped in June died by July, i.e., within 1 month. It is possible that death was associated with the combined action of higher temperatures and ultraviolet rays, which are especially intense in the mountains. This may be the reason that oocysts on mountain pastures die even in the winter, when the temperature does not exceed optimum. If there are no sheep on these pastures from October through June, the soil of the pasture should be completely free of oocysts. Sheep driven to such pastures in the summer did not become infected with coccida there (Krylov, 1960).

Such studies of survival of oocysts on various types of pastures and under rigid external environmental conditions enable us to note definite periods of pasture change as a valuable measure in the prophylaxis of cocciodiosis.

### EFFECTS OF RADIATION ON OOCYSTS

#### Effects of Ultraviolet Rays

Fish (1931a and b) noted that ultraviolet irradiation killed oocysts. Later, Litver (1935a and c, 1938a and b) studied the effects of ultraviolet light on rabbit coccidia even more thoroughly. When irradiated for 3 to 60 minutes with a mercury-quartz lamp, some unsporulated oocysts of *E. perforans* failed to develop and died. After a 3-minute exposure, 4% of the oocysts failed to develop; a 60-minute exposure killed 93% (Fig. 44). As the dosage was increased, there was a rise in the percentage of undeveloped oocysts. When the oocysts of *E. perforans* were irradiated in feces within 12 hours after emerging from the rabbit, different results were obtained (Fig. 44). At low levels (under 12 minutes) a comparatively low percentage of oocysts (4 to 15%) failed to develop. Increasing the dosage did not raise the death rate. The same occurred with *E. stiedae* oocysts. After 60-minute exposure, only 33% of

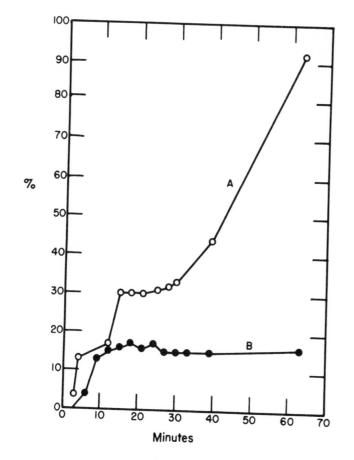

Fig. 44. Percentage of death of *E. perforans* oocysts at various stages of development under the influence of the same dose of ultraviolet rays (After Litver, 1935a). *a,* Irradiation within 10 to 15 minutes after emergence of the oocysts from the rectum; *b,* irradiation within 12 hours.

the oocysts failed to develop, demonstrating that the oocysts of this species are more resistant than those of *E. perforans*. That oocysts of *E. perforans* were less sensitive to ultraviolet irradiation 12 hours after emerging than were newly emerged oocysts seems to indicate that during sporulation there are definite stages which have a greater or lesser sensitivity to harmful types of light rays.

The most sensitive oocysts were those of *E. stiedae* at 16 hours after emerging into the external environment. At a temperature of 18 C, sporulation in *E. stiedae* ended in 72 to 74 hours. At the 16th hour, after a 3-minute dose of radiation, 20% of the oocysts stopped developing, and within 60 minutes this number had risen to 55%. The least sensitive oocysts came at the 66th hour after the start of sporulation, i.e., when the sporocysts had already formed. The 15-minute, like the 60-minute exposure, led to the death of only 8% of the oocysts. At the 71st hour, the sensitivity of the oocysts rose sharply (Fig. 45). Thus, there were two peaks of sensitivity—at the 16th and 71st hours. During the 1st days of sporulation, sensitivity increased and reached the maximum by the 16th hour, but after this it gradually decreased. Within 16 hours, division of

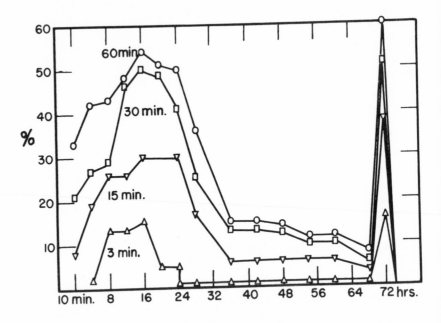

FIG. 45. Percentage of death of *E. stiedae* oocysts after ultraviolet rays at doses of 3, 15, 30, and 60 minutes at various stages of sporulation (After Litver, 1935a).

the nuclei began and the four small protrusions appeared on the spherical zygote. Until the 71st hour, structural reorganization of the sporoblasts proceeded without nuclear division. At the 71st hour, division of the nuclei began anew and the sporozoites were formed. Thus, the greatest sensitivity of *E. stiedae* to ultraviolet light coincided with the periods of nuclear division (Litver, 1935a, 1938a).

The sensitivity of *E. stiedae* oocysts did not change in relation to temperatures between $-1$ and 20 C prior to irradiation. Keeping the oocysts at temperatures of $-1$ to 20 C for 3 hours had no effect on sensitivity to ultraviolet rays. However, if the oocysts were kept at 26 C prior to irradiation, the death rate from radiation increased when compared with the control.

The temperature regime after irradiation is of special significance to the repair of oocysts after they have been damaged by ultraviolet light. At a temperature of 14 to 26 C following a 4-minute irradiation, 86 to 90% of the oocysts died, and in a few oocysts, sporulation ended within 62 to 88 hours.

Different results were obtained when the oocysts were placed at a temperature of 8 C after irradiation for a period lasting as long as 18 days. In this case, sporulation did not occur. After 18 days the oocysts were placed in an incubator at 27 C. Only 43% of the oocysts died as a result of the 4-minute irradiation, but if the oocysts were placed in a low-temperature environment for only 24 hours after irradiation, the mortality rate was 89%. Litver (1938b) hypothesized that during the state of rest at low temperatures, the best conditions existed for repair of the cytoplasm and the nucleus which allowed a high percentage of oocysts to undergo restoration of development.

The same was observed when oocysts were placed in anaerobic conditions for various lengths of time after irradiation. Twenty-four hours in such surroundings caused a 94% death rate after 4 minutes of irradiation, but 18 days of such conditions decreased it to 46%. Longer storage in an anaerobic environment raised the death rate higher because of the negative effects of anaerobiosis (Litver, 1938a). It may therefore be assumed that the oocysts of rabbit coccidia are more sensitive to ultraviolet rays than to chemical processes.

An important factor for the survival of oocysts to ultraviolet rays was the thickness of the water layer above the oocysts and the density of the fecal suspension. The thicker the layer of water and the denser the suspension of feces, the lower the number of oocysts that are killed by the same dose of light. Irradiation for 60 minutes at a distance of 100 cm in dry rabbit feces led to the death of no more than 50% of the oocysts, but all oocysts which had been cleaned were killed by the same exposure. Sporulated oocysts that were irradiated at a distance of 50 cm for 1 hour completely lost their ability to infect. No rabbits inoculated with such oocysts became infected, but the controls died of severe coccidiosis when infected with the same dose of non-irradiated oocysts (Litver, 1938b). Apparently, the death of the oocysts can be explained by the ultaviolet light, and by the heating which accompanies it. Oocysts of *E. zurnii* kept in water at 30 C died in 8 hours after exposure to sunlight (Marquardt, 1957), and the oocysts of *E. acervulina, E. tenella,* and *E. maxima* died even faster (Farr and Wehr, 1949).

### Effects of X-rays

Oocysts of *E. tenella* did not die after a dose of 3500 r (Albanese and Smetana, 1937; Waxler, 1941), and sporulated oocysts of *E. bovis* did not die from a dose of 60,000 r; such oocysts can cause an infection. However, the clinical picture of the illness was poorly expressed under these conditions (Fitzgerald, 1965).

Unsporulated oocysts of *E. magna* have retained their viability and sporulated after dosages up to 185,000 and even 370,000 r, but the treated oocysts began sporulation 1 day later than controls. In addition, about 70% of the oocysts developed in various experiments. A dose of 720,000 r caused sharp disruption in the first and second metagamic divisions. Many oocysts did not begin development, but those in which the first division occurred degenerated without forming sporoblasts. No oocysts completed sporulation.

Sporulated oocysts of *E. tenella* were irradiated with 45,000 r, after which they were used to infect 12-day-old chicks. A light infection resulted, as compared with the controls. This was probably because a dose of 45,000 r was lethal to many oocysts. Unsporulated oocysts are apparently less sensitive to ionizing radiation than sporulated oocysts.

### EFFECTS OF COCCIDIOSTATIC PREPARATIONS
### ON OOCYST SPORULATION

In the treatment of coccidiosis of domestic animals, various coccidiostatic drugs are widely employed; their action is directed at the asexual generations of the coccidia (Levine, 1961a; Horton-Smith and Long, 1963). It has been noted that these drugs do not affect the sexual stages of development. It is difficult to determine whether they have any effect on the oocysts being formed or on their subsequent sporulation, since this aspect of the question has not received much attention.

Kogan (1960) studied the effect of sulfathiazole and phthalylsulfathiazole on the sporulation of *E. necatrix* oocysts obtained from chicks which had been fed these drugs with their food. Oocysts of *E. necatrix* which had been isolated from the chicks receiving sulfathiazole for several days sporulated very poorly, and many oocysts did not form sporocysts; only 10% completed sporulation. Oocysts of *E. necatrix* obtained after 1 day's dosage of sulfathiazole sporulated slightly less than the controls. In the experiment, about 58 to 73% of these oocysts, and 72 to 82% of the controls sporulated. If sulfathiazole was given 3 to 5 days before elimination of the oocysts, sporulation was rather strongly suppressed. Sporozoites developed in 2.3% of the oocysts as compared with 41% in the control. The effects of phthalylsulfathiazole were not so distinct. The drug was given to infected chicks 3 to 6 days before the oocysts were due to appear. As a result, 32% sporulated, compared with 55% of the controls. Apparently, both drugs had a certain effect on the formation of oocysts or on the macrogametes, causing some destruction of zygotes. Ohkubo, Ikeda, and Tsunoda (1955) thought that these drugs disrupted division of the zygote nucleus.

# CHAPTER VI

# Conditions Determining the
# Infection of the Host
# by Coccidia

The life cycle of coccidia is maintained by the ingestion of sporulated oocysts from the external environment, as a result of which the host becomes infected with coccidia. This process consists of several consecutive stages. The first stage is entry of the oocyst into the host. The second stage is the excystation of the sporozoites from the oocysts in the host's intestine, and the third stage is the passage of the free sporozoites from the lumen of the intestine into the intestinal cells where subsequent endogenous development of the coccidia takes place. The development of each of these three stages requires certain conditions, the lack of which hinders infection of the host.

## ENTRY OF THE OOCYSTS INTO THE HOST

Regardless of which organ the endogenous stages are to be located in, the oocysts always enter the host passively per os. The animal ingests the oocysts along with food from the external environment where the oocysts arrived with the feces, urine, or sexual products (for example, in fishes infected by several species of coccidia which localize in the testes).

It is not entirely clear how fish are infected by the coccidium *E. gadi*. The endogenous development of this parasite occurs in the swim bladder of cods. The oocysts cluster in the swim bladder and sporulate there. It is sometimes possible to find empty sporocysts and free sporozoites in the lumen of the swim bladder. Fiebiger (1913) thought these sporozoites penetrated into the epithelium of the swim bladder, then into the connective tissue where endogenous development took place. Because the swim bladder has no connection with the esophagus, the oocysts cannot enter the external environment. It might be assumed that the elimination of oocysts into the water takes place only after the death of the fish, and oocysts are swallowed directly from the water by other fish. It is also possible that the fish become infected by eating other fish which have many oocysts in their swim bladders. Fiebiger also hypothesized that sporozoites from the cavity of the swim bladder penetrate into the connective tissue of the swim bladder wall and from there into the capillaries of the vascular system. In this case transmission to a new host may take place, as in *Schellackia,* by way of a bloodsucking vector (perhaps some sort of bloodsucking copepod).

In most cases, oocysts can be spread in the external environment where suitable hosts which can be infected with coccidia live. For example, around feeding troughs and drinking bowls of bird roosts and on the soil of runways contaminated with bird feces, one can find numerous oocysts which under favorable conditions sporulate and become a source of infection for the birds. The birds ingest these oocysts when pecking food from the soil surface.

Rabbits ingest oocysts along with contaminated hay, oats, and other food. The less often rabbit hutches are cleaned, the greater the possibility that rabbits will become infected. Cattle and sheep ingest oocysts on pasture when they eat grass containing feces with oocysts or when they drink water from ponds where the animals congregate. Repeated studies of the soil where sheep congregate have shown the presence of large numbers of oocysts (Krylov, 1956, 1960).

The dose of ingested oocysts may vary. Sometimes only isolated oocysts are ingested, but sometimes as when chickens scratch up the feces of infected birds, a large number of oocysts are taken in at one time.

Some domestic animals, such as rabbits, or even chickens, ducks, and turkeys, are heavily infected with coccidia. Because these animals tend to congretate in large groups on their separate parts of the farm, almost herdwide infections may be found, and it is then very difficult to find an uninfected animal. In addition, it is very hard to raise a coccidia-free individual. Several investigators have succeeded in raising coccidia-free animals. Tyzzer (1929), Hall (1934), Roudabush (1937), Kheysin (1940, 1947b), and others raised coccidia-free chickens, rabbits, and rats, and Krylov (1960) obtained coccidia-free lambs and kids. In all cases, it was impossible to raise coccidia-free animals over a long period of time, and spontaneous infections always resulted. Coccidia-free chickens, young rabbits, and rats have been maintained for 3 months; lambs and goats remained free 2.5 months. Because oocysts are widespread in the external environment, they can easily get into the food of the host.


## EXCYSTATION OF SPOROZOITES

Infection from ingested oocysts occurs whenever excystation takes place in the proper part of the digestive tract. This includes not only the intestinal coccida, but also the hepatic and renal varieties. Excystation always takes place in the digestive tract after which free sporozoites reach the organ where their endogenous development ultimately takes place.

The part of the digestive tract in which excystation takes place and the speed with which the process is brought about are the first questions which must be examined in order to determine the mechanism by which the host becomes infected. Secondly, the factors and circumstances that bring about excystation in the intestine of the host need to be determined.


### Site of Excystation in the Digestive Tract

In various animals and among the various species of coccidia, excystation takes place in different parts of the digestive tract. This is largely determined by the structure of the digestive tract and the functions of the individual sections. It might be expected that excystation of bird coccidia will occur in one place, while in mammals it will occur in another. The same may be said of the other species of coccidian hosts such as reptiles, fishes, amphibians, and certain invertebrates. The sites of excystation have been studied in the most detail in chickens, rabbits, rats, sheep, and cattle (Andrews, 1930; Goodrich, 1944; Itagaki, 1954; Ikeda, 1955a and b, 1956a and b; Itagaki and Tsubokura, 1958; Jackson, 1962; and others).

There is a great deal of information on where excystation occurs in chicken coccidia, but some data are contradictory. The digestive tract begins with the esophagus, which in many birds has an enlarged area, the crop. The esophagus leads into the glandular, primary stomach or proventriculus which contains many digestive glands. This stomach passes into the second or muscular stomach, the ventriculus. In birds which feed on vegetable matter, especially those which feed on grains (chickens and pigeons), this stomach is well developed and is lined with a hard wall or cuticle. The middle part of the intestine (the small intestine) forms many loops. The large intestine is separated from the small intestine by a pair of ceca. The large intestine opens into the cloaca.

Goodrich (1944) fed oocysts of *E. tenella* to 17-day-old chicks which were killed in 3 hours. Along with the intact oocysts, she found many ruptured oocysts and free sporozoites in the muscular stomach. She also observed ruptured oocysts in the small intestine and free sporozoites in the ceca. Itagaki (1954) and Pratt (1937) found sporocysts and free sporozoites of *E. tenella* in the crop and glandular stomach of chickens soon after giving them a large dose of sporulated oocysts.

More detailed studies by Doran and Farr (1962a and b) have shown that excystation of *E. acervulina* sporozoites does not occur in the crop, glandular stomach, or muscular stomach. Within 25 to 30 minutes after introduction of sporulated oocysts into the crop, the authors observed many free sporocysts with inactive sporozoites in the muscular stomach. They found free sporozoites in the duodenum immediately posterior to the muscular stomach and further along the path of the small intestine past the entrance of the bile duct. In one of four experiments, Doran and Farr found free sporocysts in the foregut. They explained this as the result of reverse peristalsis and removal from the duodenum, which may occur when a chick is drugged with ether. Apparently, the same phenomenon explains the observations of Goodrich, Itagaki and Pratt, all of whom found sporozoites and sporocysts in the crop and foregut. Doran and Farr concluded that the sporozoites of *E. acervulina* do not excyst in the upper parts of the digestive tract, i.e., in the muscular stomach and above.

Further investigations by Doran and Farr with other species of coccidia confirmed these observations. Oocysts of *E. meleagrimitis* were introduced into the crop of young turkeys, and within 15 minutes several sporocysts were found in this organ. However, the same number of sporocysts were also found in the suspension of sporulated oocysts which had been given to the turkeys. This showed that the primary site of excystation was not in the crop. At 15 to 20 minutes after introduction into the muscular stomach, a large number of sporocysts were found. Approximately 84% had been freed in the muscular stomach. The same was observed after introduction of the oocysts of *E. gallopavonis* and *E. tenella* into turkeys and chickens. These species are located in the ceca. Seventy percent of the *E. gallopavonis* sporocysts and 8.6% of *E. tenella* were released in the muscular stomach. However, no sporozoites were found in the muscular stomach.

Free sporozoites of *E. acervulina* and *E. meleagrimitis,* the endogenous stages of which are in the duodenum, were observed within 15 to 20 minutes in this part of the digestive tract. Within 30 to 60 minutes they were especially numerous in the jejunum below the point of entry of the bile duct. Sporocysts with active sporozoites inside were also found here. Only isolated sporozoites were found in the lower part of the small intestine. They were even in the ceca where they apparently drifted accidentally.

Free sporozoites of *E. gallopavonis* and *E. tenella,* the endogenous stages of which are in the ceca, were never found in the duodenum or in the upper parts of the small intestine. These areas contained large numbers of sporocysts containing inactive sporozoites. The greatest number of free sporozoites of these species of coccidia were observed in the large intestine 2 hours after infection. Sporozoites were also found in the ceca. Thus, the beginning of excystation, i.e., emergence of the sporocysts from the oocysts, occurs in the muscular stomach, but completion of this process (emergence of sporozoites) occurs in the small intestine. Apparently, the muscular stomach breaks down the oocysts mechanically, so that the emergence of the sporocysts is made easier. Sporozoites cannot emerge from intact oocysts. However, Ikeda (1956a) found experimentally that if the crop and muscular stomach are surgically removed from a chicken, infection still occurs, and excystation of *E. tenella* then takes place in the duodenum. Hence, the muscular stomach and its mechanical action on the oocysts is not absolutely necessary for infection of the chick by this parasite.

The site of excystation has not been definitely established in rabbits. However, the process does not occur in the stomach (Smetana, 1933a). Within 15 to 20 minutes after introduction per os of a large dose of sporulated *E. intestinalis* oocysts, intact oocysts were observed in the duodenum. Within 30 minutes, free sporozoites were found in the upper part of the small intestine and were later found along the entire length of the intestine.

After per os introduction, excystation of *E. nieschulzi* and *E. miyairii* sporozoites was discovered for 60 minutes along the entire small intestine of rats (Andrews, 1930).

Excystation of *E. bovis* sporozoites and other species of cattle coccidia occurs in the lower part of the small intestine, and also in the cecum during the 24 hours after introduction of the oocysts into the abomasum. Emergence of sporocysts with inactive sporozoites was also observed in this same part of the intestine (Hammond, McCowin, and Shupe, 1954; Nyberg and Hammond, 1964).

## Conditions for in vivo Excystation

Although many investigators have attempted to discover what factors determine the emergence of sporozoites in the intestine and the mechanism of this process, none has so far been able to give a final answer to such questions. Naturally, it should be assumed that the process of excystation is governed by the direct action of the digestive enzymes of the host's intestine on the oocyst, or at least that the enzymes stimulate some mechanisms within the oocyst.

Infection of rabbits and rats by various species of coccidia has been successful in all cases in which intact, sporulated oocysts were introduced directly into the duodenum and the stomach was bypassed. This has likewise proven that the actions of pepsin are not necessary for bringing about excystation.

It was shown above that infection of chickens was successful even when the oocysts were placed in the duodenum after bypassing the crop and the muscular stomach (Ikeda, 1956a). Therefore, excystation of chicken coccidia requires not so much a mechanical treatment of the oocysts in the crop and the muscular stomach as the action of enzymes found in the digestive fluid. In the opinion of Ikeda and Doran and Farr, the upper part of the digestive tract of birds is not a necessary link in the chain of processes needed for infection of chickens by *E. tenella* and other species of coccidia.

The significance of digestive enzymes in the process of excystation has been demonstrated by several very cleverly designed experiments. In chickens a ligature was placed around the pancreatic duct to stop the flow of pancreatic juice. Such chickens had been given sporulated oocysts of *E. tenella, E. mitis, E. maxima, E. necatrix,* and others per os and then observed further. No chicken treated in this manner suffered an infection (Levine, 1942a, b, and c; Ohkubo et al., 1955; Ikeda, 1955a and b, 1956a and b). This proved that the pancreatic enzymes are an important factor in determining infection of the host by coccidia and excystation of the sporozoites.

Ikeda's experiments also revealed that tying off the bile flow in chickens does not protect them against infection. Apparently, bile is not as necessary for excystation as is pancreatic juice. In infecting chicks with coccidia, the role of pancreatic juice was also confirmed by other experiments designed by Ikeda. If chickens which have had their pancreatic ducts ligated were infected with sporulated oocysts of *E. tenella,* and at the same time an extract of pancreatic juice from chickens, rabbits, or guinea pigs was introduced into the crop, infection always took place, but in control chicks with the pancreas tied off there was no infection. Analogous results were obtained when extracts of pancreatic juice were introduced into the duodenum of ligated chicks 15 to 20 minutes after the introduction of oocysts into the crop. Infections also developed when the ligated chicks were given doses of commercially manufactured trypsin directly into the duodenum (Ikeda, 1955a and b, 1956a).

Thus, a lack of pancreatic juice in the intestine of chicks hindered excystation of the sporozoites. At 2 to 6 hours after infection, the feces of ligated chicks contained 19 times more intact oocysts than were in the controls. Hence, when there was a lack of pancreatic enzymes in the intestine, a large number of oocysts did not undergo any change and were carried out of the host. It could be assumed that cutting off the flow of the pancreas led to some kind of physiological change in the mucosa of the ceca, as a result of which infection could not take place. However, if the same ligatured chicks were given a direct dose of second-generation merozoites of *E. tenella* in the ceca, further development of gamonts and the formation of oocysts occurred normally. The susceptibility of the ceca to endogenous stages during a period of pancreatic cut-off did not change as compared with controls. These data show in a sufficiently convincing manner that the proteolytic enzyme of the pancreas plays an important role in the excystation of sporozoites of bird coccidia and apparently, also in mammalian coccidia.

Excystation of the oocysts of cattle coccidia occurs in the small intestine after passage through the complex stomach. Hammond et al. (1954) and Nyberg and Hammond (1964) designed special experiments to determine if the enzymes of the stomach were absolutely necessary for excystation or whether only the enzymes in the intestine were needed for this process. Oocysts of *E. bovis* were introduced into the abomasum or its wall. This caused a serious infection in the animal. Within 24 hours, examination of the abomasal wall revealed ruptured oocysts and free sporocysts, but no free sporozoites were found. Sporocysts were seen only in the mucus, which was obtained by scraping the small intestine and cecum. Sporocysts were not seen in sections of the intestinal mucosa tissue. Hence, in order to excyst, oocysts of *E. bovis* must pass through the rumen and the other parts of the stomach [sic].

Nyberg and Hammond (1964) attempted to determine under what circumstances excystation of oocysts from cattle occurs in vivo. A fistula was made in the upper small intestine of several calves, through which oocysts of *E. bovis* were introduced in a dialysis bag along with a 0.02 M solution of cysteine hydrochloride. Excystation did not occur after 48 hours, but within 24 hours after introduction into the fistula, the oocyst walls had become thinner in the region of the micropyle and several free sporocysts were found. If such oocysts with altered micropyles were mixed with 0.5% trypsin, 0.5% steapsin, and 5% bile and again placed in the fistula, 75% of the oocysts excysted within 4 hours. Without preliminary introduction into the fistula, only isolated oocysts in the mixture of trypsin, steapsin, and bile excysted in the first 24 hours after re-entry into the fistula. Excystation did not occur if the oocysts were mixed only with trypsin or only with bile and then placed in the fistula.

The question arises whether the enzymes of the digestive tract (in particular the enzymes of the pancreas) are the only factors in excystation, or whether this process can be brought about under other conditions as well. Itagaki and Tsubokura (1958) reported that pancreatic juice alone did not cause excystation of the sporozoites and infection of the host. In their experiments, chicks which had their pancreatic ducts tied off before being given sporulated oocysts of *E. tenella* became infected and died of cecal coccidiosis. These data contradict those of Levine and Ikeda.

The intestines of some animals, such as the invertebrates, do not contain pancreatic juice, but they are nevertheless infected with coccidia, hence sporozoites undergo excystation there. *Eimeria schubergi, Barrouxia schneideri,* and other coccidia develop in the intestine of *Lithobius*. It is possible that proteolytic enzymes other than trypsin bring about the process of excystation.

To determine this some interesting experiments with rat coccidia have been carried out (Landers, 1960). A suspension of *E. nieschulzi* oocysts was cleaned of detritus and introduced into the veins of coccida-free rats. The rats had been carefully protected against the possibility of spontaneous infection. Within 30 to 60 minutes the animals were necropsied and their lungs examined in detail. Empty oocysts were found in sections of this organ, indicating that sporozoite excystation had taken place. However, no one has ever obtained proof of in vitro excystation in fresh blood.

Landers (1960) attempted to infect rats by injecting 10,000 oocysts of *E. nieschulzi* directly into the muscle and peritoneal cavity. Within 7.5 days all rats had oocysts in their feces. The number was considerably lower than in the controls which received the same dose of sporulated oocysts per os. If the experiments were properly conducted and spontaneous infection did not actually occur, then it can be concluded that excystation did take place outside the intestine and without the partici-

pation of pancreatic enzymes. Empty oocysts or other evidence of excystation were not observed in the muscles or in the peritoneal cavity.
Landers hypothesized that such enzymes as cathepsin from the skeletal
muscles and plasminogen from the blood serum may cause excystation,
but there is still no direct evidence of their participation in the process.
It is difficult to determine whether the proteolytic enzymes affect the
oocyst walls and cause their digestion. The important thing is that not
only proteins, but also polysaccharides, enter into the make-up of the
oocyst walls, and their hydrolysis must take place before excystation.

Davies and Joyner (1962) and Sharma and Reid (1962) conducted
similar experiments with chicken coccidia. The oocysts of *E. maxima, E.
acervulina, E. necatrix,* and *E. tenella* were administered in large numbers to chicks intramuscularly, intraperitoneally, and intravenously.
Within slightly more than the normal prepatent period, the birds which
received intraperitoneal injections began passing oocysts. The experiments were not always successful. In some cases (after the injection of
the oocysts of *E. acervulina* into the musculature or intraperitoneal
injection of *E. maxima*) infection did not occur. In one series of experiments, the oocysts of *E. tenella* were injected subcutaneously, as a result
of which infection occurred. Apparently, the sporozoites excysted outside the intestine and then migrated into the wall of the intestinal
mucosa. The observations of Van Doorninck and Becker (1957) and
Challey and Burns (1959) on the possibility of sporozoite transport by
macrophages give us a basis for assuming that this route may be used by
excysted sporozoites in moving from the muscles, blood, and other tissues into the intestinal mucosa.

Fitzgerald (1962) successfully infected calves with oocysts of *E. bovis*
by injecting them intraperitoneally. As a result of such injections, the
calves developed an immunity to repeated infections administered per
os. At the same time, he failed to infect calves by injecting oocysts subcutaneously, intramuscularly, and intravenously.

Thus, all the above experiments give us a basis for saying that not
only the action of the digestive enzymes, but also other proteolytic
enzymes, bring about excystation of the oocysts.

### Conditions for in vitro Excystation

Numerous in vitro experiments on sporozoite excystation have not
thus far yielded any answer to the question of the importance of the
various enzymes involved in this process. Until recent years, very few
investigators had succeeded in producing in vitro excystation of sporo-

zoites from intact oocysts in the presence of the digestive enzymes. It was partly for this reason that Landers (1960) doubted the necessity of digestive enzymes for excystation. Several authors observed excystation of sporozoites from intact oocysts of rabbit coccidia some time ago (Metzner, 1903; Reich, 1913; Waworuntu, 1924; and others). Smetana (1933a) observed excystation of small numbers of *E. stiedae* oocysts in pancreatic juice at a temperature of 37 C and explained that the activating agent for excystation was trypsin. More recently, Koyama (1956) also found that *E. stiedae* and other species from rabbits excysted in a medium which contained pancreatic enzymes.

Under the influence of pancreatic preparations, excystation of *E. tenella* sporozoites was observed by Ikeda (1960a). He used commercial trypsin, pancreatin, and duodenal juice, which had been activated by enterokinase, and all these preparations caused excystation. When oocysts of *E. tenella* were placed in the contents of the crop and the muscular stomach, they did not undergo excystation. Neither did excystation take place when the oocysts were placed in intestinal juice obtained from chicks with their pancreatic flow cut off. Preliminary treatment of oocysts with gastric juices did not raise the percentage of excystation when trypsin was later added. Ikeda thought that trypsin was the specific enzyme required for in vitro excystation. Analogous data were obtained by Ohkubo et al. (1955).

In opposition to the above data, several investigators have tried unsuccessfully to cause in vitro excystation in intact oocysts. Goodrich (1944), Itagaki (1954), and Pratt (1937) were unsuccessful in obtaining excystation of intact oocysts of *E. stiedae* and *E. tenella* by using various tissue extracts, trypsin, and other commercial preparations of digestive enzymes as well as additional bile. Landers (1960) failed to obtain excystation of intact oocysts of *E. nieschulzi* treated with pepsin, trypsin, pancreatin, lipase, and bile for 30 minutes to 24 hours. Also, excystation did not take place after increased pressure, a vacuum, and a change in the oxidation-reduction potentials of the oocysts for 30 minutes. Lotze and Leek (1960) attempted to excyst sporozoites from intact oocysts of sheep coccidia by treating them first with bile and then pancreatic enzymes. Negative results were obtained. Hammond et al. (1946) attempted to excyst sporozoites of *E. bovis* in vitro by using trypsin alone or with bile.

Doran and Farr (1962a and b) also failed to cause excystation of undamaged oocysts of *E. acervulina* in the presence of trypsin and other enzymes. They treated oocysts for 5 hours, (37 to 40 C) with trypsin in various concentrations and at a pH of 7.8, but no sporozoites emerged.

When bile was added to the trypsin no change was observed. The oocysts were treated with saliva and pepsin, then with a combination of trypsin and bile, but this did not cause excystation. Excystation was observed, however, if the oocyst wall was mechanically ruptured. Such has been also observed in *E. nieschulzi* (Landers, 1960), *E. acervulina* (Doran and Farr, 1962a and b), *E. bovis* (Hammond et al., 1946) and others.

It is highly probable that the authors who observed in vitro excystation from intact oocysts had actually used partially ruptured oocysts. The suspension may have contained free sporocysts, and the sporozoites were freed from these. Without paying any attention to this fact, the authors concluded that excystation from intact oocysts had occurred.

When the sporocysts had been mechanically freed from the oocysts, completely different results were obtained in the presence of various enzymes. This was found particularly in experiments with oocysts of chicken and turkey coccidia (Doran and Farr, 1962a and b; Farr and Doran, 1962a and b). Trypsin alone [without bile] at 1:300 (one part trypsin to 300 parts casein at a pH of 7.8 and at 40 C for 1 hour) caused very little excystation of the sporozoites of *E. acervulina, E. tenella, E. meleagrimitis,* and *E. gallopavonis.* In 1% trypsin (1: 300) at a pH of 7.4, the sporozoites underwent excystation in 30.6% of the sporocysts of *E. acervulina* in 5 hours. In 0.25% trypsin and in the same amount of time, 7.8 to 27.2% of the other four species excysted. If the sporocysts were first placed in a solution of bile for 1 hour and then 2.5% trypsin (at 1: 300) was added, the sporozoites emerged from 85 to 90% of the sporocysts within 1 hour. The reverse procedure, i.e., putting the oocysts first in trypsin for 1 hour and then adding bile caused little excystation, and not more than 50% of the sporozoites emerged. The excysted sporozoites often appeared to agglutinate and form clumps. Treatment of the sporocysts with bile alone did not cause excystation or activation of the sporozoites within the sporocysts.

The optimum pH for excystation in undiluted bile and 0.25% trypsin was 7.6. The optimum concentration of trypsin varied within wide limits from 0.025 to 1.0%, and of bile from 5 to 25%. Best excystation took place at a temperature of 37 to 41 C. At room temperature (18 to 20 C), activation and excystation of sporozoites proceeded slowly and were sometimes completely halted, especially if the sporocysts were cooled to room temperature after being at 37 to 41 C (Goodrich, 1944). However, Smetana (1933b) observed that excystation of *E. stiedae* continued even at temperatures lower than 37 C.

Excystation of *E. acervulina* may also occur in the presence of steapsin (lipase) and crystallized trypsin in combination with 5% bile. Without bile these preparations cause little excystation. Invertase and trypsin did not aid excystation, regardless of whether they were used with or without bile (Doran and Farr, 1962a and b).

Excystation of the sporozoites of chicken and turkey coccidia is stimulated by combining the action not only of trypsin and chicken bile, but also bovine bile. The percentage of excystation is practically identical in either case. For example, in some experiments, 94% of the sporocysts of *E. acervulina* excysted within 1 hour at 37 C when a mixture of 0.25% trypsin (at 1: 300) and 5% chicken bile was used. When bovine bile was used the figure was 88.1%. In another experiment the respective figures were 93.1 and 96.5%. Approximately the same results were obtained with *E. tenella.* Excystation was observed in 35.2% of the sporocysts within 3 hours when a mixture of trypsin and chicken bile was used, and in 34.2% when bovine bile was used. Under the same conditions *E. meleagrimitis* had excystation rates of 93.1 and 96.5%, respectively.

Some surface active substances, such as Tergitol and Tween-80 (polyoxyethylene sorbitan monolaurate) plus trypsin, cause excystation, but the effect is somewhat less than when bile is used. The sporozoites die rather quickly after they emerge from the sporocysts.

Bile can be replaced by sodium taurocholate. A mixture of 1% sodium taurocholate and 0.25% trypsin acts almost the same as bile. In trypsin, 3.2% of the sporozoites of *E. acervulina* excysted, but the addition of sodium taurocholate increased the percentage to 84.7. The same was observed in *E. tenella,* where the respective rates of excystation in 1 hour were 1.9 and 2.3%, and in 3 hours 13.4 and 32.9%. The same results were obtained with *E. meleagrimitis* and *E. gallopavonis* (Doran and Farr, 1962a and b; Farr and Doran, 1962a and b). Itagaki (1954) unsuccessfully attempted to obtain excystation of *E. tenella* by adding only sodium taurocholate.

Thus, it may be said that the combined presence of trypsin and bile plays an important role in the in vitro excystation of the sporozoites of bird coccidia. It is highly probable that it acts on the surface of the Stieda body, and alters the protein structure of the Stieda body which covers the sporocyst micropyle.

In the last few years several investigators have succeeded in causing excystation of sporozoites in intact oocysts of sheep and cattle coccidia (Jackson, 1962; Nyberg and Hammond, 1964). Jackson attempted to excyst sheep coccidia with extracts from the rumen, abomasum, and small intestine. The percentage of intact oocysts that excysted was very low. A larger percentage was obtained when the oocysts were introduced into the rumen in a cellophane bag and then incubated at 39 C in a solution containing trypsin and bile. The author hypothesized that excystation first required preliminary stimulation of the oocysts by some kind of substance found in the rumen. Preliminary treatment with an extract obtained from the abomasum did not increase excystation.

Jackson established that the necessary stimulus for excystation in vitro was saturation of the environment containing the sporulated intact oocysts with carbon dioxide. It was important to create anaerobic surroundings in order to prepare the oocysts for the subsequent action of trypsin and bile. If the environment of the oocysts contained an isotonic buffer, stimulation of excystation caused the sporozoites only to emerge from the sporocysts; they would not leave the oocysts. If the buffer was diluted with water, the sporozoites left the oocysts through the micropyle. The best results in causing in vitro excystation were obtained when the oocysts were incubated in water (not in potassium bichromate). Apparently, bichromate decreased the penetrability of the wall for enzymes which play the role of secondary stimulators for excystation. It may be that the micropyle became dense and the sporozoites could not leave the oocysts.

The higher the gaseous concentration of $CO_2$ in the environment, the greater was the percentage of oocysts which ultimately excysted. When the environment was saturated with 15 to 20% $CO_2$ mixed with $N_2$ in a bicarbonate buffer at 38 C, subsequent excystation in trypsin and bile was observed in 47 to 64% of the oocysts. In 30 to 40% $CO_2 + N_2$, about 90% excysted. This process could occur at temperatures from 30 to 43 C. At 46 C no sporozoites could be obtained. Several substances such as sodium dithionite, cysteine, and ascorbic acid had a stimulating effect on the oocysts when $CO_2$ and $N_2$ were present, and at the same time the percentage of subsequent excystation was increased. In 20% $CO_2$ plus nitrogen, it was necessary to activate the oocysts for only 6 hours in order for the highest percentage of excystation to occur.

Thus, it seems that a similar stimulation of oocysts occurs in the rumen of sheep when they are prepared with digestive enzymes for secondary activity which ultimately leads to excystation in the intestine. However, the experiments of Hammond et al. (1954) and Nyberg and Hammond (1964) have shown that such preliminary action of the rumen contents in vivo is not at all necessary for subsequent excystation of cattle coccidian oocysts.

In experiments in vitro, the oocysts of *E. arloingi* treated with primary stimuli were placed in an environment consisting of 0.25% trypsin and 10% bile with a phosphate buffer at a pH of 7.4 and a temperature of 39 C. The sporozoites became motile within the sporocysts, and after the disappearance of the Stieda body they emerged into the cavity of the oocyst. Here they moved, and, using their forward ends, contacted the micropyle of the oocyst. After the oocysts were treated with water or a buffer mixture, the sporozoites passed through the micropyle to the outside. This process was completed in 30 minutes to several hours. Excystation occurred only when the oocysts were first acted upon by bile and then were placed in trypsin (Jackson, 1962). Trypsin could be replaced by papain. Taurocholate and Tween-80 worked in a manner similar to bile.

Excystation in vitro occurs in two stages: (1) the action of the primary stimulators on the intact oocysts which prepares the oocyst for excystation, and (2) the action of the secondary stimulators which causes the actual excystation. Apparently, the primary stimulators act both in the rumen and in the small intestine, while the secondary ones act only in the small intestine.

Rupturing of the oocysts prior to excystation is not needed, and, apparently, a change in hydrostatic pressure in the intestine is a more important factor in the stimulation of excystation than is mechanical rupturing of the oocysts. Treatment of the oocysts with hypertonic solutions also raises the percentage of excystation. What probably happens in this case is that the permeability of the oocyst wall changes, or the enzyme systems of the oocysts are activated, thus affecting their internal wall and the wall of the sporocysts.

Jackson's data on excystation of sporozoites from intact oocysts were confirmed by Nyberg and Hammond (1964) with three species of cattle coccidia. If the sporocysts were freed from the oocysts of *E. bovis, E. ellipsoidalis,* and *E. auburnensis,* excystation occured when they were incubated in a mixture of 0.5% trypsin, 0.5% steapsin and 5% bovine bile at a pH of 7.5 and a temperature of 37 C. The onset of excystation was observed within 20 minutes, and within 1 hour 90% of the sporocysts yielded sporozoites. However, excystation of intact oocysts occurred only after the preliminary action of an aqueous solution of 0.02 M cysteine saturated with a gaseous mixture of $CO_2$ and air or $CO_2$ and $N_2$ at a temperature of 37 C. After such treatment, the oocysts were subjected to the mixture of bile and enzymes, which led to excystation. After an 8-hour exposure at a temperature of 37 C in a mixture of air and $N_2$ (40: 60) and an acetate or phosphate buffer solution (or even without the buffer solution) a 4-hour incubation in a mixture of bile and enzymes did not lead to excystation. Negative results were also obtained with the use of a mixture of $CO_2 + N_2$ (40: 60) in an acetate buffer. On the other hand, a mixture of $CO_2 + N_2$ (40: 60) in a phosphate buffer activated the oocysts, and excystation was observed in 25% of the oocysts within 4 hours. A mixture of $CO_2$ and $N_2$ in an unbuffered medium activated the oocysts in the same amount of time, and 75% of the *E. bovis* oocysts underwent excystation. Oocysts treated in this fashion had an altered micropyle which was more delicate than the normal type.

After treatment of the oocysts with 0.02 M cysteine in a phosphate buffer, the internal wall of the oocysts became more delicate. It was able to stain violet in toluidine blue. This color was not observed in untreated oocysts. If the percentage of $CO_2$ in the gas mixture was increased to 60, the percentage of excystation was increased. Treatment with such a gas mixture for 2 to 4 hours led to the excystation of 37 to 41% of the oocysts, and within 14 hours, as many as 90% underwent excystation. Oocysts placed in 0.02 M cysteine during preliminary treatment with $CO_2$ + $N_2$ (50: 50) for 10 hours later excysted rapidly in a mixture of trypsin, lipase, and bile. Incubation in the enzymes for 1 hour led to 26% excystation, but 6 hours increased the figure to 93%. If the amount of bile was reduced from 5 to 1%, 1.5 times fewer oocysts excysted. Under identical conditions, the oocysts of *E. bovis* underwent a greater amount of excystation than those of *E. ellipsoidalis*. After 6 hours of incubation, 95% of the former species excysted, while the second had only 80% free.

In the comparison of the excystation of *E. bovis* and *E. arloingi,* the former had maximum excystation after 10 hours of treatment in 50% $CO_2$, while the latter needed only a 4-hour treatment in 40% $CO_2$. A small percentage of excystation was obtained in vitro after preliminary treatment of the oocysts of sheep coccidia with saliva and 1% steapsin. The same was observed by Nyberg and Hammond (1964) for *E. bovis.*

From the above data, a certain clarity has come about in recent years in our understanding of the processes of in vivo and in vitro excystation. Apparently, the factors determining this process in vivo are similar in birds and mammals. In any case, there is no doubt that the basic role in excystation belongs to trypsin in combination with bile. It may be that other factors have a supplementary influence on this process, but we still have no experimental proof of this. It is more or less clear that in birds and mammals, the anterior part of the digestive tract (esophagus, crop, and stomach) does not play a part in the activation for excystation. It remains unexplained what factors cause excystation in the intestines of invertebrates which have no trypsin. It is highly probable that this enzyme is not the only one which causes the emergence of sporozoites from sporocysts. However, in the intestines of birds and mammals, excystation and infection cannot occur without trypsin.

As far as in vitro excystation is concerned, intact oocysts can be made to yield sporozoites only in sheep and cattle coccidia. For activation of this process, it is necessary first of all to create anaerobic conditions and to alter the oocyst micropyle. After this preliminary treatment, excystation begins in a mixture of bile, trypsin, and steapsin, or in bile and trypsin. It is difficult to say whether the same factors will have an activating effect on the coccidia of other animals. Further experiments are needed.

## Mechanism of Excystation

Under the influence of pancreatic enzymes, the sporozoites inside the sporocysts begin to move and exit through the micropyle which is covered by the Stieda body. This was observed by Ikeda (1960a) in *E. tenella* and by Nyberg and Hammond (1964) in cattle coccidia. The Stieda body swells and then seems to dissolve. The sporozoites quickly emerge through the narrow opening of the micropyle where the Stieda body formerly was and remain within the oocyst wall. The sporocyst residuum is usually destroyed during this process. The sporozoites swim around in the cavity of the oocyst and emerge through the oocyst micropyle. The oocyst micropyle becomes somewhat more delicate after treatment with $CO_2$ and cysteine, which facilitates the emergence of the sporozoites. During passage through the micropyle, the sporozoites shorten their bodies and move by spinning around their longitudinal axes. The sporozoites of *E. acervulina* emerge in a similar manner (Doran and Farr, 1962a and b). The sporozoites of *E. acervulina* became active shortly after the destruction of the Stieda body, but in *E. meleagrimitis* and *E. tenella* this was noted prior to destruction.

Thus, in vitro, the sporozoites from the intact oocysts first leave the sporocyst then the oocyst. In those cases in which mechanical rupturing of the oocyst is observed, the sporocysts are not within the oocyst wall and the sporozoites emerge from the sporocysts. This is apparently what happens in the intestines of birds after the oocysts are mechanically ruptured in the muscular stomach.

Preliminary treatment of oocysts causes a change in the permeability of the oocyst walls. Staining of the internal wall of such oocysts with toluidine blue serves as an index of this phenomenon. Because excystation from the sporocysts is possible without preliminary treatment and requires only the action of intestinal enzymes, the preliminary treatment must affect only the oocyst wall and not the sporocysts. Some investigators are of the opinion that the mechanism of excystation is associated with activation of internal enzymes in the oocyst which act on the wall from inside to cause a change in the structure of the wall. It can be assumed that the intestinal enzymes or other factors cause activation of these internal enzymes and at the same time other conditions for excystation arise.

**PENETRATION OF SPOROZOITES INTO THE
TISSUES OF THE HOST**

After the sporozoites emerge from the oocyst, they begin to penetrate into the mucous membrane of the host's intestine. There are very few data on how this process occurs. It is most often noted that the sporozoites penetrate directly into the epithelial cells where further development takes place. However, data exist which show that the sporozoites of chicken coccidia, regardless of their ultimate location as schizonts in various parts of the epithelial cells of the intestinal mucosa, first migrate into the connective tissue of the mucosa. In the opinion of some investigators who have studied this process in *E. brunetti* and *E. praecox,* such sporozoites penetrated into the tunica propria and did not undergo any further development and possibly died (Boles and Becker, 1954; Tyzzer et al., 1932). Tyzzer et al. also observed that large numbers of sporozoites of *E. necatrix* accumulated in the connective tissue of the crypts approximately 1 hour after infection. They were later found in the epithelium of the crypts.

Other investigators reported that the sporozoites of *E. tenella* (Gill and Ray, 1957; Pattillo, 1959; Challey and Burns, 1959), *E. necatrix* (van Doorninck and Becker, 1957), *E. acervulina* (Wagner, 1965; Doran, 1966a and b), *E. gallopavonis* (Farr, 1964), and *E. meleagrimitis* (Clarkson, 1959a) passed through the epithelium into the connective tissue of the crypts where preparation for further development within the epithelial cells took place. Greven (1953), for example, found early stages of development of *E. tenella* in the fibroblasts of the intestinal mucosa. Van Doorninck and Becker (1957) reported that the sporozoites of *E. necatrix* appeared to be in the macrophages before beginning development in the epithelium of the crypts. In sections from the wall of a chick's intestine that was fixed within 12 hours after infection, it was possible to see the sporozoites penetrating the epithelium of the villi in the middle portion of the small intestine. Within 18 hours many sporozoites were observed in the macrophages. At this time there were no initial stages of development in the epithelium. Thus, the authors suggested that first the sporozoites passed through the epithelium into the connective tissue in the direction of the muscular layer, where they were ingested by macrophages. Most of the sporozoites were probably digested by these cells, but some macrophages containing sporozoites migrated toward the crypts where the sporozoites left the macrophages and penetrated into the epithelial cells. Hence, these cells were infected by the sporozoites not from the lumen of the intestine, but from the direction of the connective tissue.

An analogous phenomenon was described for *E. tenella* which develops in the ceca. Challey and Burns (1959) found sporozoites in the epithelium (mainly in the basal part of the cells) 2 to 4 hours after infection. At later periods no sporozoites were observed in the epithelium, but there were many of them in the macrophages of the connective tissue. Within 12 hours, the sporozoites were again found in the epithelial cells of the glands. Apparently, they penetrated there from the macrophages which were destroyed after the sporozoites emerged from them. The same phenomenon was recently observed in sporozoites of *E. acervulina* (Doran, 1966a and b). They migrated either through the epithelial cells or else between them into the connective tissue. The first sporozoites in the epithelial cells of the duodenum were observed within 10 minutes after infection. Within 1 hour, up to 92% of the sporozoites concentrated in the epithelium and only isolated ones were found in the connective tissue. Within 3 hours, the number of sporozoites in the epithelium had decreased to 66 to 75%, but their number in the connective tissue had risen to 35%. However, within 6 hours 89% had penetrated the tunica propria. Here they were ingested by the macrophages and could be observed within the latter for several days. The first sporozoites in the epithelium of the crypts were found within 18 hours. At first they were under the nuclei of the epithelial cells, but then moved and began settling above the nuclei. Sporozoites probably began development in the macrophages and then moved to the epithelium where schizogony took place.

It is very difficult to explain this strange phenomenon in which sporozoites are transported by macrophages. It remains unclear whether this occurs in all cases or whether only a certain part of the sporozoites are subject to the attack of macrophages. The possibility is not eliminated that macrophages and sporozoites may enter the bloodstream and thence go into the internal organs. Development does not occur in the bloodstream.

The question as to the routes by which sporozoites move from the intestine into the liver has not been entirely answered. *Eimeria stiedae* develops in the epithelium of the bile ducts in rabbits. The sporozoites of this species have been observed in the upper part of the intestine. Smetana (1933b and c) explained that the sporozoites penetrate the liver through the bloodstream, not the bile duct. The endogenous stages of *E. stiedae* have developed in the livers of rabbits in which the bile duct had been ligated. When sporozoites were introduced intravenously, the rabbits were infected by the liver coccidium *E. stiedae,* and sporozoites were found in the blood vessels of the liver after large doses of sporulated oocysts had been introduced into the intestine. However, no sporozoites were found in the lymph vessels. Apparently, sporozoites used the blood system to penetrate not only the liver but also the kidneys, but this question needs further study and clarification. Pellérdy (1965) discussed the possibility that sporozoites of *E. truncata* penetrated the kidneys from the intestine.

# CHAPTER VII

# Species Properties of Coccidia
# of Domestic Animals

This chapter contains short descriptions of the coccidia of domestic animals. A description of the oocysts and endogenous stages of development is given if such are known. Detailed information on these coccidia can be found in the books by Davies, Joyner and Kendall (1963) and Pellérdy (1965).

## COCCIDIA OF MAMMALS

### Coccidia of Cattle

At the present time approximately 12 to 15 species of coccidia are known, but some of them are apparently not valid species and have been described only on the basis of their oocysts.

*Eimeria alabamensis* Christensen, 1941. Oocysts are pyriform, ellipsoid or pyramidal, colorless, and 13–24 by 11–16 microns (mean 18.9 by 13.4 microns). A micropyle is not visible. Sporulation occurs in 4 or 5 days at 20 C. Sporocysts are 19.5 by 6.5 microns. No sporocyst or oocyst residuum is present.

Endogenous development occurs in the small intestine. Schizonts develop in the nuclei of epithelial cells in the apical part of the villi. Gamonts are located in the lower part of the small intestine, but in severe infections they may be found in the cecum or upper part of the large intestine. Schizonts are present from the 2nd through the 8th day. Merozoites, measuring 7–9 by 1.4–2.1 microns, develop in schizonts which contain 15 to 32 merozoites. The first oocysts appear in the intestine on the 6th day after inoculation. The prepatent period is 6 to 11 days (mean 8.6), and the patent period may be up to 10 days (mean 7.2 days).

Found in North America.

*Eimeria bovis* Fiebiger, 1912. The oocysts are ovoid, light brown or yellow, and a micropyle is present at the narrow end. Oocysts are 23–24 by 17–23 microns (mean 27.7 by 20.3 microns). Sporulation takes place in 2 to 3 days.

First-generation schizonts are in the lower part of the small intestine, and second-generation schizonts and gamonts are found in the large intestine and cecum. First-generation schizonts measure up to 300 microns and contain 100,000 merozoites which are 11 microns long. The schizonts are in the endothelium of lymph vessels of the villi and mature in 14 to 18 days after infection. Second-generation schizonts are about 10 microns, and contain 30 to 36 merozoites which appear 2 days after maturation of the first-generation schizonts. Second-generation schizonts are located in epithelial cells. The gamonts mature in epithelial cells within 3 days. The prepatent period is 18 to 21 days, and the patent period lasts 5 to 7 days, but sometimes may last up to 14 days.

This species has worldwide distribution and has been found in the USSR.

*Eimeria auburnensis* Christensen and Porter, 1939. The oocysts are ovoid or ellipsoid. The micropyle is a light colored, thin zone at the narrow end of the oocyst. The surface of the oocyst wall is smooth or sometimes slightly rough and is yellowish brown. Oocysts are 32–46 by 20–25 microns (mean 38.4 by 23.1 microns). Sporulation occurs in 2 to 3 days.

Endogenous development takes place in connective tissue cells of the villi in the lower part of the small intestine. Small numbers of macrogamonts are found 2 to 6 m from the ileocaecal valve on the 14th to 18th day after infection. The microgametocytes are very large (85.1 by 65.1 microns) and easily visible to the naked eye. They are found on the 18th day after infection. According to Hammond et al. (1961), the prepatent period is 18 to 20 days. Oocysts are passed for 2 to 7 days.

Found in Europe, North and South America.

*Eimeria bukidnonensis* Tubangui, 1931. Oocysts are pyriform or ovoid, slightly blunt on one end, and have a broad micropyle at the narrow end. The oocyst wall is yellowish brown and has a delicate striping. Dimensions (according to various authors) are 33.3–50.4 by

26.2–33.7 microns (mean 36.6–48.6 by 26.7–35.4 microns). According to Christensen (1941), sporulation occurs in 4 to 7 days, according to Lee (1954) in 17 days, and according to Baker (1938) in 24 to 27 days.

All endogenous stages are in the small intestine. Merozoites have been observed on the 13th day after infection, and oocysts are found in the lower part of the small intestine on the 25th day.

Found in the USSR, Europe, North America, Brazil, Nigeria, and the Philippine Islands.

*Eimeria ellipsoidalis* Becker and Frye, 1929. Oocysts are ellipsoid, or sometimes cylindrical or spherical. A micropyle is not visible. The wall is colorless. Oocysts are 20–26 by 13–17 microns (mean 23.4 by 15.9 microns). Sporulation occurs in 2 or 3 days.

All endogenous stages are located in the epithelium of the crypts of the small intestine. Schizonts are 10.6 by 9.4 microns and contain 24 to 36 merozoites which are 8 to 11 microns long. Gamonts and oocysts are in the lower part of the small intestine. The prepatent period is 10 days.

Found in the USSR, Europe, North and Central America.

*Eimeria zuerni* (syn., *E. zurnii*) Martin, 1909. The oocysts are spherical, subspherical, or sometimes ellipsoid. The wall is smooth, colorless, and has no micropyle. Oocysts (according to various authors) are 12.2–28.7 by 12.2–22 microns (mean 17.1–20.9 by 14.6–15.6 microns). Sporulation occurs in 2 or 3 days.

All endogenous stages are in the cecum or large intestine, but microgametocytes are sometimes found in the lower part of the small intestine. Schizogony occurs up to the 19th day. Mature schizonts are 13.2 microns in diameter and contain 24 to 36 merozoites which are up to 12 microns long. The number of asexual generations is unknown. Microgametocytes measure 13.5 by 10.6 microns and are found on the 12th to 19th days of the infection. The first oocysts appear in the tissues on the 12th day. The prepatent period is 12 to 18 days.

Found in the USSR, Europe, North America, Brazil, and the Philippine Islands.

*Eimeria brasiliensis* Torres and Ramos, 1939. Oocysts are ovoid and have a micropyle covered with a micropylar cap 10 to 12 microns wide and 2 to 4 microns high. The oocyst wall is thickened toward the micropylar end. The oocysts are colorless and measure 33.7–49.0 by 21.1–33.2 microns (mean 37.5 by 27 microns). The sporocysts have an elongate ovoid shape and measure 16–21.8 by 7.2–8.7 microns. Sporulation takes up to 12 to 14 days.

Found in the USSR, Europe, North America, Brazil, and Nigeria.

*Eimeria canadensis* Bruce, 1921. Oocysts are usually ellipsoidal or cylindrical, have a micropyle and are 28–37 by 20–27 microns (mean 32.5 by 23.4 microns). The oocyst wall is smooth and slightly yellowish brown. Sporulation time is 3 or 4 days.

Found in the USSR and North America.

*Eimeria cylindrica* Wilson, 1931. The oocysts are almost cylindrical, colorless, and 16–27 by 12–15 microns (mean 23.3 by 13.3 microns). A micropyle is absent. Sporulation time is 2 days.

Found in North America, India, and Europe.

*Eimeria mundaragi* Hiregaudar, 1956. Oocysts are ovoid and 36–38 by 25–28 microns. A 0.5-micron wide micropyle is present on the narrow end. The oocyst wall is smooth, yellowish, and slightly thickened in the region of the micropyle. Sporocysts are ovoid, 14.8 by 9.1 microns, and have a residual body. Sporulation occurs in 1 or 2 days.

Found in India.

*Eimeria pellita* Supperer, 1952. Oocysts are ovoid and have a micropyle at the narrow end. The oocyst wall is dark brown and slightly roughened. Oocysts measure 36.2–40.9 by 26.5–30.2 microns. Sporulation occurs in 10 to 12 days. Sporocysts are 14–18 by 6–8 microns and contain a large residual body which is 5 to 7 microns in diameter.

Found in Austria.

*Eimeria subspherica* Christensen, 1941. The oocysts are spherical, subspherical, or sometimes ellipsoid, translucent and colorless, without micropyles, and 9–13 by 8–12 microns (mean 11 by 10.4 microns). Sporulation takes 4 to 5 days. Sporocysts are spindle-shaped and have no residual body.

Found in North America.

*Eimeria wyomingensis* Huizinga and Winger, 1942. Oocysts are almost ovoid or pyriform, have a micropyle at the narow end and 37–44.9 by 26.4–30.8 microns (mean 40.3 by 28.1 microns). The wall is roughened and yellowish brown. Sporocysts are spindle-shaped and 19 microns long. A sporocyst residuum is present. Sporulation takes 5 to 7 days.

Found in North America.

Other species (*Eimeria boehmi* Supperer, 1952; *E. bombayansis* Rao and Hiregaudar, 1930; *E. ildefonsoi* Torres and Ramos, 1939; *E. khurodensis* Rao and Hiregaudar, 1953; *E. thianethi* Gwéléssiany, 1935; *Isospora aksaica* Basanoff, 1952) which have been described from cattle are probably not valid and should therefore be listed as synonyms of other species.

### Coccidia of Sheep and Goats

Sheep and goats harbor several species of coccidia, the oocysts of which are not structurally distinguishable from one another. However, experimental cross infections (Shiyanov, 1954; Krylov, 1959b; Tsygankov et al., 1959) have demonstrated that coccidia of sheep and goats are highly specific and probably are separate species. However, Deiana and Delitala (1953) regarded cross-infection as possible. Since this question has not been answered for sure and demands further investigations, I introduce below a description of the species under their generally accepted names.

*Eimeria arloingi* (Marotel, 1905) Martin, 1909. Oocysts are elongate ellipsoid or ovoid and have a distinct micropyle and micropylar cap. The oocyst wall appears colorless to brown, with the internal layer colored. According to Krylov (1959b), oocysts are 20.9–31.9 by 16.5–23.1 microns (mean 27.2 by 18.8 microns). The sporocysts are ovoid, have a residual body and measure 13 by 6 microns. A polar granule is present (Fig. 46). The sporulation time is 2 or 3 days.

All endogenous stages are located in the small intestine. Schizonts develop in the endothelium of vessels of the villi. Mature schizonts appear on the 13th to 21st day and reach 150 microns in diameter. They contain hundreds of thousands of merozoites. Gamonts develop in the epithelial cells and can first be found in the intestine on the 19th day. Microgametocytes are polycentric and form several residual bodies. The prepatent period is 18 to 20 days, and the patent period is 6 or 7 days.

Found in the USSR, Europe, North America and Tunisia.

The possibility has not been eliminated that the large schizonts of this species were earlier described as *Globidium gilruthi* Chatton, 1910. At that time it was assumed that the so-called globidial schizonts and large oocysts of *E. intricata* belonged to one species, *G. gilruthi;* at the present this question has not been resolved.

*Eimeria faurei* (Moussu and Marotel, 1902) Martin, 1909. Oocysts are ovoid and have a distinct micropyle 2 to 3 microns in diameter which does not have a polar cap. The oocyst wall is smooth and yellowish; the outer layer is half as thick as the inner one. According to Krylov (1959b), the oocysts are 20.9–36.3 by 16.5–27.5 microns (mean 30.95 by 22.59 microns). Sporocysts are 14–16 by 8–9 microns and have a residual body (Fig. 46). Sporulation lasts up to 4 days.

Details of endogenous development are unknown. It is known that large schizonts up to 100 microns in diameter with thousands of merozoites have been seen in the small intestine. The prepatent period is 15 to

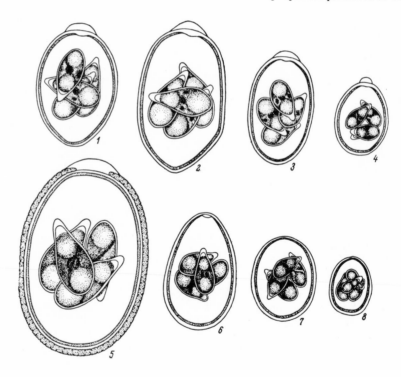

FIG. 46. Oocysts of sheep and goat coccidia (After Krylov, 1960). *1, E. granulosa; 2, E. ahsata; 3, E. arloingi; 4, E. crandallis; 5, E. intricata; 6, E. faurei; 7, E. ninakohlyakimovae; 8, E. parva.*

16 days, and the patent period is 6 to 7 days.

Found in the USSR and has a worldwide distribution.

*Eimeria ninakohlyakimovae* Yakimoff and Rastegaieff, 1930. Oocysts are short ellipsoids or ovoid and have no micropyle (but a micropyle can sometimes be noted) [sic]. No polar cap is present. The oocyst wall is thin, translucent, and light yellow. According to Krylov (1959b), oocysts are 16.5–27.5 by 14.3–23.1 microns (mean 22.2 by 18.08 microns). Sporocysts are ovoid, 5–6.5 by 3–4.5 microns and have a residual body (Fig. 46). Sporulation takes place in 1 or 2 days.

Endogenous development takes place in the small and large intestine. Schizonts develop in the small intestine and reach a diameter of 30 microns. From 40 to 200 merozoites are formed within the schizonts. Gamonts are in the epithelium of the large intestine.

Found in the USSR, Europe, North America, and Tunisia.

*Eimeria parva* Kotlán, Mócsy and Vajda, 1929. Oocysts are spherical or subspherical, and a micropyle and polar cap are absent (Fig. 46).

According to Krylov (1959b), oocysts are 9.9–18.7 by 7.7–14.3 microns (mean 14.3 by 12.34 microns). Sporulation lasts up to 4 to 5 days.

Endogenous development occurs in the small intestine, where large schizonts up to 250 microns in length and 128 microns in width are found. They develop in 12 to 14 days. Two types of schizonts have been reported. In one type the nuclei are scattered about the entire cytoplasm; in the other, they are situated along the periphery. It is possible that the merozoites from the first type yield microgametocytes, and those of the second type yield macrogametes. The schizonts penetrate the connective tissue of the villi. Gamonts develop in the cecum and the large intestine. The prepatent period lasts 14 to 15 days (Krylov, 1959b).

Found in the USSR, Europe, North America and Tunisia.

*Eimeria ahsata* Honess, 1942. Oocysts are elongate ellipsoid or ovoid and have a micropyle and polar cap. Oocysts sometimes show individual granules of a residual body. In addition, there is a polar granule. Oocysts are 29–33.5 by 22–25 microns (mean 32.7 by 23.7 microns). Sporocysts are 18–20 by 7–10 microns and contain a residual body (Fig. 46). The prepatent period is 20 to 21 days.

Found in the USSR and North America.

*Eimeria christenseni* Levine, Ivens and Fritz, 1962. Oocysts are ovoid and slightly flattened at one end. The micropyle is covered with a cap 6 to 10 microns wide and 1 to 3 microns high. After sporulation one or several polar granules are visible. Oocysts measure 34–41 by 23–28 microns (mean 38 by 25 microns). The external wall is 1 micron thick, and the internal wall is brownish and 0.4 microns thick (Fig. 47).

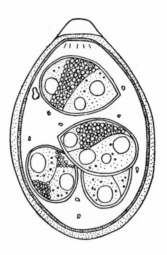

FIG. 47. *E. christenseni* (After Levine et al., 1962).

Endogenous development takes place in the epithelium of the villi of the small intestine where gamonts and small schizonts localize. Large schizonts, with 100,000 merozoites which measure 280 by 150 microns, develop in endothelial cells of the lymph vessels of the villi. These cells are ruptured as the schizonts mature, and they remain in the lumen of the vessels. Merozoites are 10 to 12 microns long and 0.8 to 1.2 microns wide. A second type of schizont, which localizes in the epithelium, are 10–14 by 9–10 microns (mean 12 by 9 microns). They form 16 to 22 merozoites. Gamonts are under the nuclei of the epithelial cells. Their dimensions are almost identical.

Found only in goats in North America.

*Eimeria crandallis* Honess, 1942. Oocysts are broadly ovoid and have a micropyle and cap that are 3.3 to 6.6 microns wide and 1.7 microns high. The oocyst wall is smooth and colorless. According to Krylov (1959b), oocysts are 16.5–29.7 by 12.1–20.9 microns (mean 22.1 by

16.85 microns). Sporocysts are 8.3–11.2 by 5.4–8 microns. Polar granules and a sporocyst residuum are present (Fig. 46). The prepatent period is 13 to 14 days.

Found in the USSR and North America.

*Eimeria granulosa* Christensen, 1938. Oocysts are urn-shaped, broadened at the micropylar end and 22–35 by 17–25 microns (mean 29.4 by 20.9 microns). A polar cap, which is 5 microns wide and 1 to 2.5 microns high, is present (Fig. 46). The oocyst wall is yellowish brown. The sporulation time is 3 to 4 days.

Found in the USSR, Germany, and North America.

*Eimeria intricata* Spiegl, 1925. These oocysts are the largest of all species, 39–53 by 27–34 microns (mean 47 by 32 microns). The shape is ellipsoid, and a micropyle 6 to 10 microns wide and polar cap 6 to 11 microns wide are present (Fig. 46). The oocyst wall is rough, brown in color and cross-striated. Henry (1932b) reported there were three walls. Sporocysts are ovoid, measure 16–18 by 8–10 microns and have a large residual body. Polar bodies are present. Sporulation takes 3 or 4 days.

Endogenous development occurs in the lower part of the small intestine. Development of gamonts also takes place in the cecum. Schizonts range in size up to 45 by 65 microns. Merozoites are as long as 12 microns. All stages are in epithelial cells. The prepatent period is 22 to 27 days, and the patent period is 4 to 11 days (Davis and Bowman, 1965).

Found in the USSR, Europe, North America, and Tunisia.

*Eimeria punctata* Landers, 1955. Oocysts are spherical or subspherical and measure 17.8–25.1 by 16.2–21.1 microns (mean 21.2 by 17.7 microns). Small pits are present on the oocyst wall. The outer oocyst wall is yellowish and the inner layer is light green. A micropyle is present on the narrow end and is covered with a flat cap. Sporulation takes about 2 days.

Found only in sheep in the USSR and North America.

Most of these species are apparently valid, but several other species of sheep and goats have been described, the validity of which is doubtful (*E. longispora* Rudovsky, 1922; *E. pallida* Christensen, 1938; *E. aemula* Yakimoff, 1931).

## Coccidia of Horses

*Eimeria leuckarti* (Flesch, 1883) Reichenow, 1940. Oocysts are very large (80–87 by 55–59 microns from the ass and 75–78 by 50–55 from the horse). They are ovoid and narrow at one end where there is a micropyle. The oocyst wall is 6 to 7 microns thick, dark brown, and has a granular surface. Sporocysts are 30–42 by 13–14 microns. Sporulation requires up to 22 days.

Schizonts and gamonts are found in the small intestine. Various authors have observed globidial schizonts measuring 300 microns in diameter. In addition, macrogametes have been found. Because no experimental research has been conducted, details of the life cycle are unknown.

Found in Europe, North America, and India.

Two other species, *E. solipedum* Gousseff, 1934 and *E. uniungulati* Gousseff, 1934 with spherical or ellipsoid oocysts, have never been observed anywhere by anyone except the author who described them. They were found in the USSR.

## Coccidia of Camels

*Eimeria noelleri* (Henry and Masson, 1932) Pellérdy, 1956. The oocysts are broadly ovoid, have a micropyle and are 80–100 by 65–80 microns, and the wall is 10 to 15 microns thick. The wall is composed of three layers; the middle layer is thick and yellow. Sporocysts are 40–50 by 14–25 microns (mean 44.7 by 17.8 microns) and have a residual body.

Endogenous development occurs in the abomasum and distal part of the ileum, where globidial schizonts up to 350 microns in diameter are found. A detailed study of the cycle has not yet been conducted.

Found in the USSR, Europe, and North America.

*Eimeria cameli* Nöller, 1922. Oocysts are spherical or subspherical and have a micropyle 5 to 6 microns wide. No polar cap is present. According to Tsygankov (1950), oocysts are 16.5–40 by 12–30.2 microns. Sporocysts are lemon-shaped, 15–17 by 10 microns and have a residual body. A polar granule is present in the oocysts. Sporulation takes 6 to 8 days.

Endogenous stages are found in the small intestine 2 m from the pylorus. Schizonts measuring 10 to 16 microns contain 20 to 24 merozoites.

Microgametocytes are up to 19 microns in diameter and have small numbers of microgametes.

Found in the USSR and Europe.

*Eimeria dromedarii* Yakimoff and Matschoulsky, 1939. The oocysts are ovoid or subspherical and 23.1–32.5 by 19–25.2 microns (mean 27.7 by 23.2 microns). There is a polar cap 6.3 to 8.4 microns wide on the one rather flattened end. The oocyst wall is brown and 1.4 microns thick. Oocyst and sporocyst residua and polar granules are absent.

Found in the USSR.

*Eimeria pellerdyi* Prasad, 1960. The oocysts are ovoid or ellipsoid, 22.5–24 by 12–13.5 microns (mean 23.2 by 12.6 microns) and have no micropyle or polar cap. Sporocysts are 9–10.5 by 4.5–6 microns and have a residual body. Sporulation takes place in 4 or 5 days.

Found in a camel at the London zoo.

*Isospora orlovi* Tsygankov, 1950 is probably not a parasite of the camel. The oocysts which Tsygankov interpreted as a new species might more likely have been transitory or accidentally carried into the feces of the camel. The oocysts of *I. orlovi* are very similar to the oocysts of *I. lacazei* from sparrows. According to Levine and Mohan (1960), the oocysts of *Isospora* found in sheep and cattle are actually *I. lacazei*. The same has been successfully demonstrated in the case of *Isospora* of chickens, ducks, and geese (Anpilogova, 1965).

## Coccidia of swine

*Eimeria debliecki* Douwes, 1921. The oocysts are ovoid or subspherical, 12.8–28.8 by 12.5–19.5 microns, and have a smooth wall with no micropyle. A polar granule is present. Sporulation takes 5 to 9 days.

Endogenous stages occur in the small intestine (Wiesenhütter, 1962; Vetterling, 1966). First-generation schizonts are 30 to 180 cm from the stomach; second-generation schizonts are up to 400 cm from the stomach; and gamonts are 650 cm along the small intestine. All stages of development are in the epithelium of the villi. However, according to Boch and Wiesenhütter (1963) the endogenous stages are found in the epithelium of the crypts, in the connective tissue, and even in the submucosa. First-generation schizonts, measuring 10.2 by 9.8 microns, form 16 merozoites which are arranged like orange sections. First-generation merozoites are 12–15 by 1.8 microns. They appear within 48 hours after infection, and the greatest numbers are observed at the 60th hour. Second-generation schizonts are 13–16 (13.2) by 10–15 microns. Mature schizonts with 32 merozoites appear within 96 hours after infection and are retained up to the 144th hour. Second-generation merozoites are 6 to 8 microns long and 1.8 microns wide. Gametogony lasts 1.5 days, and within 120 hours after infection young macrogametes and microgametocytes, measuring 9–14 by 7–9 microns, are found. Their number increases by the 144th hour. The microgametocytes form up to 75 microgametes which are 5 to 6 microns long. The prepatent period is 146 hours (6.5 days), and the patent period lasts 6 to 7 days.

Found in the USSR, Europe, North and South America, the Island of Java, and the Congo.

*Eimeria scabra* Henry, 1931. Oocysts are ellipsoid or ovoid, 22.4–35.6 by 16–25.5 microns and have a rough wall. A distinct micropyle is present on the narrow end. A polar body is present. The sporocysts are ellipsoid and measure 16–19.2 by 6–4 microns. Sporulation takes 9 to 12 days.

Endogenous development has been poorly studied. Schizonts and gamonts are found in the mucosa of the large intestine. The prepatent period lasts 9 to 10 days, and the patent period is 4 to 6 days.

Found in the USSR, Europe, and North America.

*Eimeria perminuta* Henry, 1931. The oocysts are usually ovoid, but rarely spherical and measure 11.2–16 by 9.6–12.8 microns. The oocyst wall is rough. Sporulation takes 11 days.

Found in the USSR and North and South America.

*Eimeria spinosa* Henry, 1931. Oocysts are ovoid or ellipsoid and measure 16–22.4 by 12.8–16 mm [sic]. The wall is brown and the surface is covered with tiny spines that are 1 micron high. Sporocysts are 9.1–11.7 by 5.2–6.5 microns and have residual bodies. Sporulation takes 10 to 21 days.

Endogenous development takes place in the small intestine. First-generation schizonts, which are 8 to 10 microns, develop in the epithelium of the villi and crypts. Merozoites are 4–6 by 1–1.5 microns. Each schizont forms approximately 20 merozoites. Microgametocytes are 6 to 8 microns in diameter; they are somewhat smaller than the macrogametes (Wiesenhütter, 1962). The prepatent period is 7 days.

Found in the USSR, North America, Europe, and Hawaii.

*Eimeria polita* Pellérdy, 1949. The oocysts are usually ellipsoid, less often ovoid, 22–31 by 17–22 microns, and the wall is yellowish brown. A micropyle is not visible. Sporulation takes 8 to 9 days. Sporocysts are 16–17 by 6 microns and have a residual body. The prepatent period lasts 8 or 9 days.

Found in Europe and North America.

*Eimeria scrofae* Galli-Valerio, 1935. The oocysts are cylindrical, flattened at one end and 24 by 15 microns. A micropyle is present. Sporocysts are 15 by 6 microns.

Found in Europe.

*Isospora suis* Biester and Murrary, 1934. The oocysts are subspherical and 20–24 by 18–21 microns. The wall is 1.5 microns thick and yellowish brown. A micropyle is not present. Sporulation takes 4 days. Sporocysts are 17.6 by 10–12 microns.

The endogenous stages of development are in the epithelium of the small intestine, and cells infected with the parasites submerge into the connective tissue. The prepatent period lasts from 6 to 8 days.

Found in North America and Europe.

*Isospora almaataensis* Pajchuk, 1951. The oocysts are ovoid or spherical, dark gray in color, and 24.6–31.9 by 23.2–29 microns (mean 27.9 by 25.9 microns). The wall is smooth and 3 microns thick. Sporocysts are 11.6–18.9 by 8.7–11.6 microns and have a residual body. Sporulation requires 5 days.

Found in the USSR.

## Coccidia of Dogs and Cats

*Eimeria canis* Wenyon, 1923. The oocysts are ovoid or ellipsoid and measure 18–45 by 11–28 microns. The external wall is easily removed. A micropyle is present. Sporulation takes 3 or 4 days.

Found in dogs in Europe, Australia, and North America.

*Eimeria felina* Nieschulz, 1924. The oocysts are ovoid or ellipsoid and 21–26 by 13–17 microns. The wall is smooth and colorless. A sporocyst residuum is present.

Found in cats in Holland.

*Eimeria cati* Yakimoff, 1933. Subspherical oocysts measure 21–25.3 by 12.6–14.7 microns and spherical oocysts are 16 to 22 microns. A polar granule is present.

Found in cats in the USSR.

*Isospora felis* (Wasielewsky, 1904) Wenyon, 1923. The oocysts are ovoid and 35–45 by 23–35 microns. No micropyle is present. Sporocysts measure 20–24 by 18–21 microns and have a large residual body. Sporulation takes 2 days.

Endogenous development occurs in the epithelium along the length of the small intestine. Schizonts are sometimes found in the large intestine and cecum. Two generations of schizonts are present. The first generation lasts from the 2nd through the 4th day. Schizonts measuring 10–16 by 4–7 microns form 40 to 60 merozoites. Second-generation schizonts appear on the 5th or 6th day and form from 2 to 24 (mean 12 to 16) merozoites which are 3 to 5 microns long. Microgametocytes are 25–47 by 19–35 microns, have one residual body, and are approximately the same size as macrogametes. Gamonts appear on the 6th day and are found up to the 8th day. They are found not only in the small intestine, but also in the cecum. The prepatent period is 7 to 8 days (Hitchcock, 1955; Lickfeld, 1959; Neméseri, 1960).

This parasite is specific for cats. It is found in the USSR, Europe, North America, and Japan.

*Isospora canis* Neméseri, 1959. The oocysts are ovoid, colorless, and measure 36–44 by 29–36 microns (mean 39 by 32 microns). Sporulation lasts 4 days. Sporocysts are ellipsoid, measure 24 by 20 microns, and have a residual body which is 5.5 microns in diameter. The prepatent period is 11 days.

This species is found only in dogs in Europe.

*Isospora rivolta* (Grassi, 1879) Wenyon, 1923. The oocysts are ovoid, 20–25 by 15–20 microns and have a distinct micropyle at the narrow end. Sporocysts are 14–16 by 9–10 microns and have a large residual body. Sporulation takes 4 days.

Endogenous development occurs in the epithelium of the small intestine. Wenyon and Sheather (1925) found schizonts and gamonts in the connective tissue of the villi. However, it is possible that the authors were dealing with a mixed infection of *I. rivolta* and *I. bigemina*.

Found in dogs and cats in the USSR, Europe, and North America.

*Isospora bigemina* (Stiles, 1891) Lühe, 1906. Oocysts are ovoid or spherical and have no micropyle. Dogs have round oocysts that are 18–20 by 14–16 microns and small ones measuring 10–16 by 7.5–10 microns. The latter are also found in cats. The small oocysts sporulate before leaving the host while still in the connective tissue of the villi of the small intestine. The large oocysts sporulate in 4 days in the external environment. It is highly probable that these are oocysts of two distinct species. The small oocysts are often eliminated with a damaged wall. Often the feces of dogs and cats contain individually mature sporocysts that measure 8–9 by 5–7 microns and have a residual body.

Endogenous development takes place in the middle portion of the small intestine. According to Wenyon (1926), development occurs in both the epithelium and the connective tissue of the villi. The schizonts in the epithelium contain 8 merozoites, while the subepithelial schizonts form 12 merozoites. The microgametocytes are larger than the macrogametes. The oocysts form in the connective tissue. In an acute infection caused by *I. bigemina,* nonsporulated oocysts may also be passed into the external environment. In dogs the prepatent period lasts 5 days, while in cats it is 6 to 7 days. The patent period lasts approximately 27 days.

Found in the USSR and has a worldwide distribution.


## Coccidia of Rabbits

Coccidia of the rabbit *(Oryctolagus cuniculus)* have been studied in detail, and apparently at least eight species of coccidia, genuinely established by experiment, are parasitic in this animal (Fig. 48).

*Eimeria stiedae* (Lindemann, 1865) Kisskalt and Hartmann, 1907. Oocysts are ovoid and have a smooth wall which is yellowish brown. A micropyle is present on the narrow end of the oocysts. The oocysts are 30–40 by 16–25 microns. A residual body is found in the oocysts after sporulation. It consists of several light-refracting granules situated between the sporocysts. A large residual body is present in the sporocysts. Sporulation takes 3 or 4 days.

Endogenous development occurs in the epithelium of the bile ducts of the liver. Schizogony is observed on the 5th to 8th day. Schizonts measure 10 to 15 microns in diameter and form somewhat thick merozoites. In addition, there are other schizonts in which many thin merozoites form. The merozoites are 5–13 by 1–2 microns. Gamonts are observed

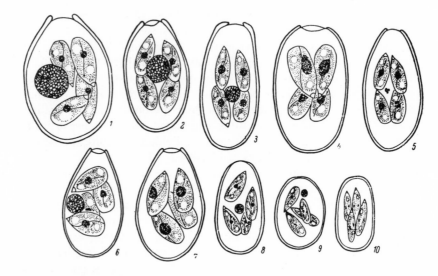

Fig. 48. Oocysts of rabbit coccidia. *1, E. magna; 2, E. media; 3, E. coecicola; 4, E. irresi-dua; 5, E. stiedae; 6, E. intestinalis; 7, E. piriformis; 8* and *9, E. perforans; 10, E. nagpurensis* (*1* to *9*, Original; *10*, After Gill and Ray, 1960).

in large numbers on the 16th day after infection. Apparently, the first gamonts appear on the 14th or 15th day after infection. Microgameto-cytes are 20 to 25 microns in diameter and have a large residual body. The prepatent period is 16 to 17 days.

Worldwide distribution includes the USSR.

*Eimeria coecicola* Cheissin, 1947. The oocysts are ovoid or cylindrical and have a micropyle on the slightly narrowed end. Oocysts are bright yellow or light brown, and measure 25.3–39.9 by 14.6–21.3 microns (mean 33.1–35.5 by 16.9–19.6 microns). The oocysts contain a residual body which is 1.3 to 6.7 microns in diameter (mean 3.9 to 5.3 microns). Sporocysts are 17 by 8 microns and have a residual body 2 to 4 microns in diameter. Sporulation lasts 3 days.

Schizogony occurs in the lower part of the small intestine, and gam-etogony takes place in the appendix and cecum. The endogenous stages are always in the epithelium, except that the oocysts are submerged in the tunica propria. Development takes place in the small intestine until the 7th day, but afterwards in the cecum and appendix. On the 6th day the schizonts are 12 to 18 microns in diameter and contain 8 to 12 merozoites. The microgametocytes, which are 20–22 by 12–17 microns, are not larger than the macrogametes. The prepatent period is 9 days and the patent period is 8 to 9 days.

Found in the USSR, Hungary, and India.

*Eimeria intestinalis* Cheissin, 1948. The oocysts are broadly pyriform or ovoid and have a distinct micropyle on the narrow end. The wall is somewhat thickened around the micropyle. The oocysts are light brown and measure 21.3–35.9 by 14.6–21.3 microns (mean 27.1–32.2 by 16.9–19.8 microns). The oocyst residuum is 2.5–9.9 microns (mean 4.9–7.5 microns) in diameter. Sporocysts are ovoid, 10 by 5 microns, and have a residual body 1.25–6.25 microns in diameter. Sporulation takes 3 days.

Endogenous development occurs in the epithelium of the villi and crypts of the lower part of the small intestine, as well as in the large intestine. Three generations of schizonts develop in the small intestine. They are found there until the 10th day. On the 8th day gametogony begins, not only in the small intestine, but also in the appendix, cecum, and ascending colon. First-generation schizonts measure 15 to 25 microns and form 15 to 60 spindle-shaped merozoites which are 7–8 by 2 microns in diameter. The second generation is found from the 5th to 10th day. At this time there are two types of schizonts: Type A schizonts are large, measure 13 to 30 microns, and contain 70 to 120 narrow, long merozoites (10–12 by 0.5–1 microns) arranged within the schizont like bunches of bananas; Type B schizonts have a diameter of 10 to 12 microns and contain 5 to 20 short, broad merozoites. From the 7th to 10th day, the third-generation schizonts are found; these are 12 to 16 microns, and contain 15 to 25 narrow merozoites. The schizonts sometimes contain a small residual body. The adult microgametocytes are oval, monocentric, and measure 22 to 27 microns; their diameter is about that of the macrogametes. The prepatent period lasts 9 to 10 days, and the patent period is no longer than 10 days.

Found in the USSR, Hungary, and India.

*Eimeria irresidua* Kessel and Jankiewicz, 1931. Oocysts are ellipsoid, slightly broadened at one end, and have a micropyle. The side walls are usually parallel. The micropyle is easily seen, and around it the wall is slightly elevated. Oocysts are a light or dark brown and measure 25.3–47.8 by 15.9–27.9 microns (mean 35.1–40.8 by 20.2–23.8 microns). No oocyst residuum is present. Sporocysts are ovoid, measure 13.3–19.1 by 6.2–9.3 microns, and have a large residual body, which is 4 to 9.3 microns in diameter or slightly elongate. Sporulation takes 3 or 4 days.

Endogenous development takes place in the middle small intestine. The schizonts are in the apical part of the epithelial cells of the villi. Only the early stages of the gamonts are located in the epithelium. These cells, with the parasites, submerge into the connective tissue, where the oocysts develop. Four generations of schizonts are apparently formed. On the 3rd day the first-generation schizonts, which are 10 to 13 microns and contain 12 to 20 merozoites, appear. The merozoites are 5.8 to 10 microns long and 1.5 to 3 microns wide. On the 4th day, the second-generation schizonts are visible (6 to 10 microns, with 5 to 15 merozoites measuring 5–8 by 1.2–2.5 microns). At the same time even larger schizonts are found with 1 to 20 merozoites that are 7 to 10 microns long. On the 5th day schizonts with 2 to 4 merozoites, with a length of 5 to 7 microns, are found. This is probably the third generation. On the 7th day schizonts 7 to 12 microns long are found which contain 12 to 25 merozoites, each 5 to 8 microns long. This is the 4th generation. From the 5th day, the early gamonts appear. The microgametocytes reach 30 to 60 microns and have many centers where gametes are formed. The microgametocytes are 3 or 4 times larger than the macrogametes. The prepatent period is 7 days, and the patent period lasts for 10 to 13 days.

Worldwide distribution includes the USSR.

*Eimeria media* Kessel, 1929. Oocysts are ovoid or ellipsoid, 18.6–33.3 by 13.3–21.3 microns (mean 26.2–30.2 by 16.7–17.5 microns) and have a distinct micropyle. There is a thickening of the external wall around the micropyle. The external wall ranges from light yellow to dark brown in color. The oocyst contains a residual body which is 2.7 to 7.9 microns. The sporocysts are ovoid, 6.6–14.6 by 5–7 microns, and have a residual body measuring 1 to 3 or 2.7 microns. Sporogony lasts from 2 to 3 days.

Endogenous development occurs in the duodenum and the upper part of the small intestine. Pellérdy and Babos (1953) have found gamonts in the large intestine and appendix. However, there is a possibility that this report really concerns a double infection of *E. media* and *E. coecicola,* the oocysts of which are very similar. The schizonts are in the epithelium of the villi above the cell nucleus. The first and second generation are situated along the entire villus, but the third generation is found at the distal end. The gamonts are located here, and they also penetrate into the connective tissue of the villi. Mature first-generation schizonts are 12 to 14 microns, contain 6 to 11 merozoites and appear on the 2nd or 3rd day. On the 4th day, the second-generation schizonts form; they are oval, 9 to 12 microns long and have 2 to 18 merozoites situated in a pattern resembling orange sections. The merozoites are 7 to 10 microns long. Third-generation schizonts measure 18 to 25 microns and form 35 to 130 merozoites, each of which is 6 to 7 microns long and 1 to 1.5 microns wide. These merozoites appear on the 4th to 6th day. Gametogony begins on the 4th or 5th day. The microgametocytes are usually monocentric and 15–25 by 9–17 microns. The prepatent period is 5 days, and the patent period lasts 6 to 8 days.

The species has a worldwide distribution including the USSR.

*Eimeria magna* Pérard, 1925. The oocysts are ovoid and have well defined micropyles around which the external wall forms a sizable cylindrical thickening. The external oocyst wall is brown. Oocysts are 26.6–41.3 by 17.3–29.3 microns (mean 32.9–37.2 by 21.5–25.5 microns). The oocyst residuum is 3.9 to 12.0 microns (mean 6.7 to 9.7 microns). The sporocysts are ovoid, measure 15 by 7.8 microns, and have a residual body measuring 3.9–4.9 microns. Sporulation takes 3 to 5 days.

Endogenous development occurs in the lower and middle part of the small intestine. The gamonts are occasionally found in the appendix and the cecum. Development of all the stages begins in epithelial cells, but the infected cells sink into the tunica propria where mature schizonts, gamonts, and oocyst localize.

Sometimes the epithelial cells infected with schizonts retain their connection with the epithelial layer and sink only slightly into the tunica propria. The mature first-generation schizonts, which appear on the 4th day, measure 11 to 17 microns and form 3 to 24 merozoites that are 7 to 11 microns long. By the end of the 5th or the start of the 6th day, second-generation merozoites, which measure 5 to 12 microns, can be seen. The schizonts measure 14 to 36 microns and form 40 or 50 merozoites. There are certain differences in the sizes of the second-generation schizonts and merozoites which are associated with sexual dimorphism. By the end of the 6th day or by the 7th day, third-generation merozoites, 5 to 10 microns long, in schizonts, 25 to 40 microns long, appear. From 25 to 80 merozoites form in each schizont. On the 7th and 8th day another generation (and perhaps even two generations) of schizonts appears. They measure from 25 to 45 microns, and each contains 30 to 60 and perhaps as many as 125 merozoites. The merozoites measure 6 by 2 microns. From the 4th day it is often possible to see polynuclear merozoites. The mature microgametocytes are observed on the 7th or 8th day. As a rule, they are polycentric and reach a size of 17 to 40 microns, i.e., approximately twice the size of the macrogametes. The prepatent period is 7 to 8 days, and the patent period lasts 15 to 19 days.

This species has a worldwide distribution which includes the USSR.

*Eimeria perforans* (Leuckart, 1879) Sluiter and Schwellengrebel, 1912. The oocysts are ovoid, cylindrical, subspherical, or even spherical and have a micropyle in the large oocysts (the micropyle is sometimes not visible in smaller oocysts). The larger the oocyst, the more easily one can see the thickening in the external wall around the micropyle. The oocysts are 13.3–30.6 by 10.6–17.3 microns (mean 20.3–25.4 by 12.4–15.3 microns) and are usually colorless. The oocyst residuum is 1.3–5.3 microns (mean 2.4–4.3 microns) in diameter. Oocysts without a residual body are seldom seen. The sporocysts measure 8–9 by 4.5 microns and have a residual body 1 to 1.5 microns in diameter. Sporulation takes 1 to 2 days.

Endogenous development occurs in the middle part of the small intestine. All stages are in the epithelium of the villi and crypts. First-generation schizonts with merozoites are found on the 3rd or 4th day. Schizonts are 13 to 14 microns and form 50 to 120 narrow merozoites which are 7–9 by 0.5–0.7 microns. In addition, there are small schizonts 5 to 10 microns, with 4 to 8 merozoites measuring 4–5 by 1 micron. The number of generations has not been determined. The microgametocytes are ovoid, 12 to 14 microns and are monocentric. They are not larger than the macrogametes. The prepatent and patent periods last 5 to 6 days.

The species has worldwide distribution including the USSR.

*E. piriformis* Kotlán and Pospesch, 1934. Oocysts are pyriform, have a well developed micropyle on the narrow end, and are 26.0–32.5 by 14.6–19.5 microns (mean 29.6–31.7 by 17.7–18.8 microns). The wall is yellowish brown. An oocyst residuum is absent. The sporocysts are ovoid and 10.5 by 6 microns. The sporocyst residuum is 2.5 to 5.5 microns in diameter. Sporulation takes 3 to 4 days.

Endogenous development occurs in the large intestine, and not in the small intestine as Pellérdy (1953, 1965) reported. All stages are usually in the ascending colon, cecum, and appendix. Endogenous stages occur in the epithelium of the crypts above the host cell nuclei. Development of first-generation schizonts lasts for 5 or 6 days; they are 20 microns in diameter and contain 15 to 25 spindle-shaped merozoites which are 15 microns long and 2 microns wide. Second-generation schizonts develop from the 7th to 8th day; they measure 11 to 14 microns and have 25 to 55 short narrow merozoites which are 6–7 by 1 micron. On the 9th and 10th day the third-generation schizonts appear; these measure 17 to 20 microns and contain 15 to 50 merozoites that are 11–12 by 1.5 microns. Asexual reproduction ends by the 11th day. The first gamonts appear on the 8th day. The microgametocytes measure 16 to 18 microns, are monocentric and are no larger than mature macrogametes. The prepatent period lasts 10 days, and the patent period is about the same length of time.

Found in the USSR, Hungary, and France.

In addition to these species, *E. matsubayashii* Tsunoda, 1952, has been described in Japan. The oocysts are broadly ovoid, 24.8 by 18.2 microns and have a micropyle. The residual body has a diameter of 6.2 microns. It is highly probable that this is not a valid species, since the oocysts are very similar to those of *E. intestinalis,* but the endogenous stages resemble those of *E. media.* A second species, *E. nagpurensis* Gill and Ray, 1960, has been found in India. The oocysts are barrel-shaped with parallel walls and measure 20–26 by 10–15 microns (mean 23 by 13 microns). No micropyle is present. The sporocysts resemble an oat grain. An oocyst residuum is absent, but a sporocyst residuum is present (Fig. 48). These oocysts have not been found in rabbits in the USSR.

*Eimeria neoleporis* Carvalho, 1942 has been described from the American rabbit, *Sylvilagus floridanus mearnsi.* The species was successfully used to infect a domestic rabbit (Carvalho, 1944). The oocysts are nearly cylindrical, have a micropyle on the slightly narrowed end and measure 32.8–44.3 by 16.7–22.8 microns (mean 38.8 by 19.8 microns) (Fig. 49). The residual body consists of 3 or 4 granules which disappear as sporulation is completed. The sporocysts are ellipsoid, measure 17.1 by 8 microns, and have a residual body. Sporulation takes 2 or 3 days.

FIG. 49. Oocyst of *E. neoleporis* (After Carvalho, 1942).

Endogenous development occurs in the cecum and appendix. On the 5th day, adult first-generation schizonts appear; these measure 20 microns and contain 40–48 merozoites. The schizonts are in the tunica propria. On the 8th day, mature second-generation schizonts appear; they contain 60 to 70 delicate merozoites which are 25.7 by 1.6 microns. On the 9th day, third-generation schizonts mature. There are two types of schizonts (small ones with an average of 14 merozoites and larger ones with 60 to 86 merozoites). The merozoites from the small schizonts measure 18 by 3.5 microns, and those from the large ones reach a size of 31.5 by 1.5 microns. Gamonts appear on the 10th to 11th day after infection. The prepatent period lasts 12 days, and the patent period is 10 days.

Pellérdy (1954b) was of the opinion that *E. coecicola* was the same species as *E. neoleporis,* but these are different species. They differ in their oocysts (residual body), the duration of the prepatent period and several properties of endogenous development (localization and development of asexual generations), and the times when gamonts appear.

*Eimeria exigua* Yakimoff, 1934, is synonymous with *E. perforans* (Kheysin, 1947b).

## COCCIDIA OF BIRDS

### Coccidia of Chickens

Nine species of the genus *Eimeria* and one species of *Wenyonella* have been described from chickens (Figs. 50–55). The validity of *Isospora gallinae* Scholtyseck, 1954 still remains open, since the possibility has not been eliminated that these are transitional oocysts of *I. lacazei* which are widespread among sparrows. Tyzzer (1929) reported *Cryptosporidium parvum,* a parasite of mice, in chickens. Levine (1961a) named the species in chickens *C. tyzzeri.*

*Eimeria acervulina* Tyzzer, 1929. The oocysts are ovoid, have a smooth wall and a micropyle which is barely noticeable at the narrow end. According to Tyzzer (1929), oocysts are 17.7–20.2 by 13.7–16.3 microns (mean 19.5 by 14.3 microns); according to Becker (1952) they are 16.4 by 12.7 microns, and according to Levine (1961a), they are 16 by 13 microns. A polar granule is present. Sporulation takes 1 day.

Endogenous development occurs in the anterior small intestine, mainly in the duodenal loop. Schizonts are in the epithelium of the villi, above the host cell nuclei (Fig. 50). On the 3rd day, mature schizonts appear with 16 to 30 merozoites which are 6 by 0.8 microns. At the beginning of the 4th day mature gamonts appear in the epithelium. The microgamonts measure 11 by 9 microns. The prepatent period is 4 days.

The species has a worldwide distribution which includes the USSR.

*Eimeria brunetti* Levine, 1942. The oocysts are ovoid and 20.7–30.3 by 18.1–24.2 microns (mean 26.8 by 21.7 microns). The wall is smooth and no micropyle is present. One or two polar bodies are formed during the 1-day period of sporulation.

Endogenous development occurs primarily in the lower part of the small intestine, the ceca, the colon, and the cloaca (only the first-generation schizonts develop more in the upper part). The schizonts and gamonts are under the nuclei of epithelial cells of the villi (Fig. 51). In severe infections the mature schizonts penetrate into the connective tissue. On the 2nd to 3rd day after infection, mature first-generation schizonts, measuring 30 by 20 microns and containing 200 merozoites, are present in the upper part of the small intestine. On the 4th day two types of schizonts are found in the middle and lower parts of the small intestine (large ones 20.9 by 16.2 microns with 50 to 60 merozoites, the small ones measuring 9.8 by 8.8 microns with 12 to 14 merozoites). These are either second- or third-generation schizonts, or second-generation schizonts with sexual differentiation. The gamonts form on the 5th day. The microgametocytes are mono- and polycentric, and are larger than the macrogametes. The prepatent period lasts 5 days, and the patent period is about 10 days.

The species apparently has a worldwide distribution.

*Eimeria hagani* Levine, 1938. Oocysts are broadly ovoid, have no micropyle and are 15.8–20.9 by 14.3–19.5 microns (mean 19.1 by 17.6 microns). After sporulation the polar body forms in about 1 or 2 days.

Endogenous development occurs in the anterior part of the small intestine. The schizonts are in the epithelium. The prepatent period is 6 days, and the patent period is as long as 8 days.

Found in North America, Europe, and India.

*Eimeria maxima* Tyzzer, 1929. Oocysts are ovoid and 21.4–42.5 by 16.5–29.8 microns (mean 30 by 20 microns). The oocyst wall is slightly roughened, and a micropyle is visible on the narrow end. A polar gran-

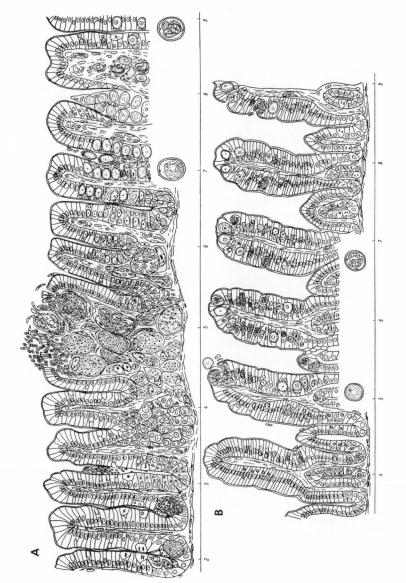

FIG. 50.  Diagram of the endogenous development of the oocysts of *E. tenella* (*A*) and *E. mitis* (*B*) (After Tyzzer, 1929).

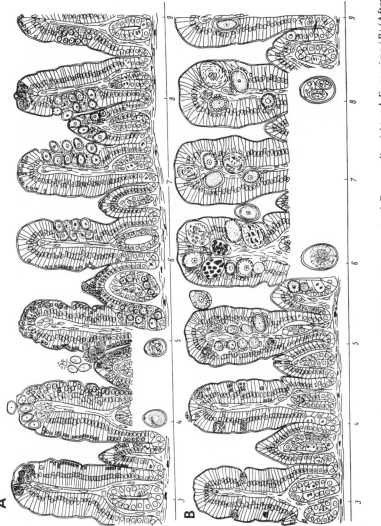

FIG. 51. Diagram of the endogenous development of the oocysts of *E. acervulina* (*A*) and *E. maxima* (*B*) (After Tyzzer, 1929).

ule is present. Sporulation lasts 2 days.

Endogenous stages are found along the entire length of the small intestine, but most are in the anterior part. The schizonts develop above the nuclei of the epithelial cells, but the gamonts are below the nuclei (Fig. 50). The epithelial cells infected with the parasite submerge into the connective tissue. The first-generation schizonts measure 10 by 8 microns and form 8 to 16 merozoites that are 7 to 9 microns long and 2 to 2.5 microns wide. The second generation is structurally similar to the first and begins within 60 hours after infection. The mature schizonts appear on the 4th or 5th day, but the first generation begins within 12 to 24 hours and matures on the 3rd or 4th day. Greyish white foci are visible on the surface of the mucosa where the schizonts cluster. The microgametocytes, which appear on the 5th day, are large (up to 39 microns) and contain 500 to 600 microgametes. The microgametocytes are larger than the macrogametes. The prepatent period is 5 days (7 days according to Scholtyseck).

The species has worldwide distribution including the USSR.

*Eimeria mitis* Tyzzer, 1929. The oocysts are subspherical, have a polar body and measure 11–19 by 10–17 microns (mean 15.6 by 13.8 microns). According to different authors the oocysts are approximately the same 15.8–16 by 14–16 microns. Sporulation takes 48 hours.

Endogenous development occurs in the anterior portion of the small intestine. The adult schizonts are in the epithelium of the villi, usually under the host cell nuclei (Fig. 52). On the 4th day, 6 to 24 (mean 16) merozoites measuring 4.5–5.5 by 1.3–1.6 microns form in the mature schizonts. The polar body is easily seen in the schizonts. Gamonts are numerous on the 5th day. The microgametocytes are monocentric and 9.5–13.5 microns long. The prepatent period lasts 5 days.

The species has worldwide distribution including the USSR.

*Eimeria mivati* Edgar and Seibold, 1964. Oocysts are broadly ovoid or ellipsoid, colorless, and have a micropyle in the form of a thinning of the wall. Oocysts are 10.7–20 by 10.1–15.3 microns (mean 15.6 by 13.4 microns). There is a polar granule near the micropyle. Sporocysts are 10.5 by 5.6 microns and have a residual body (Fig. 53). Sporulation lasts 18 to 21 hours.

Endogenous development occurs along the entire length of the small intestine and in the rectum. All stages are in the epithelium of the villi or crypts (Fig. 54). First-generation schizonts develop in the epithelium of the crypts of the upper third of the small intestine. Within 36 hours

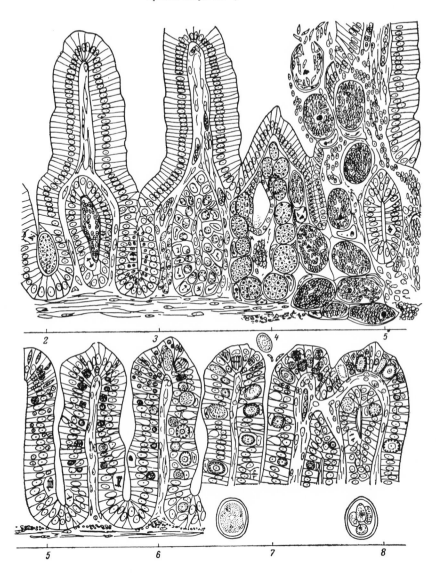

FIG. 52. Diagram of the endogenous development and oocyst of *E. necatrix* (After Tyzzer et al., 1932).

mature schizonts, measuring 9.0–11.7 by 9–11 microns (mean 9.6 by 2.3 microns) with 10 to 30 merozoites, are present. Second-generation schizonts develop within 55 to 65 hours and are under the nuclei of the epi-

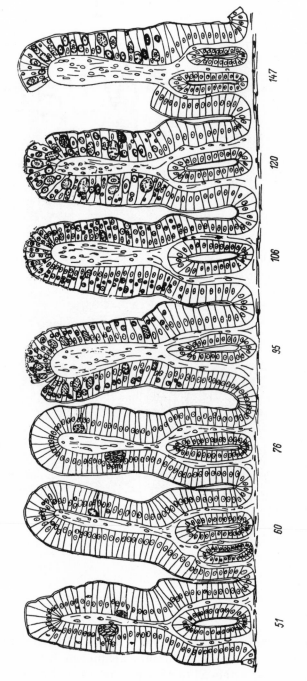

FIG. 53. Diagram of the endogenous development of *E. brunetti* (after Boles and Becker, 1954).

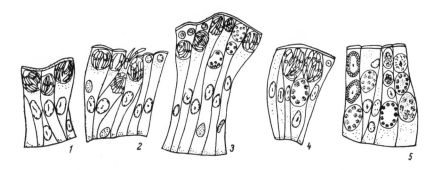

FIG. 54. Endogenous development of *E. mivati* (After Edgar and Seibold, 1964). *1*, Adult first-generation schizonts at 37 hours; *2*, adult second-generation schizonts at 66 hours; *3*, adult third-generation schizonts at 80 hours; *4*, adult fourth-generation schizonts at the 96th hour; *5*, macrogametes and adult microgametocyte in the duodenal epithelial cells at the 120th hour.

thelial cells. Their dimensions average 9.2 by 7.2 microns, and they contain 16 to 20 merozoites which measure 6.5 by 1.5 microns. They develop around the residual body, which has a diameter of 7 or 8 microns. Many second-generation schizonts are found not only in the upper regions of the intestine, but also in the lower part of the small intestine, the ceca, and the colon. The third-generation schizonts appear within 80 hours. They are in the epithelium of the villi along the entire intestine, but they are often found in the anterior part. The schizonts are 6.6 by 5.0 microns and form 10 to 14 merozoites which are 5.8 by 1.1 microns. The residual body is 5 to 6 microns. Fourth-generation schizonts appear within 93 to 96 hours; these measure 12.9 by 11.2 microns and have 16 to 20 merozoites that are 7.5–9.6 by 0.8–1.4 microns (mean 9.1 by 1.2 microns). There is a residual body (10 to 11 microns long) in the schizonts. The microgametocytes measure 14 by 11.7 microns and are the same size as the macrogametes. They form slightly over 100 microgametes. The prepatent period lasts 93 to 96 hours, and the patent period is as long as 12 days.

Found in North America and Europe.

*Eimeria necatrix* Johnson, 1930. The oocysts are ovoid, have no micropyle and are very similar to those of *E. tenella.* Oocysts measure 13–22.7 by 11.3–18.3 microns (mean 16.7 by 14.2 microns) (Tyzzer et al., 1932). According to Becker (1952), the oocysts are larger (19.3 by 16.5 microns). Edgar (1955) and Davies (1956) reported even larger oocysts (20.4 by 17.2 and 20.5 by 16.8 microns), respectively. The oocysts are sometimes ovoid. A polar body is present. Sporulation lasts 2 days.

Endogenous development occurs in the small intestine and ceca. Schizogony takes place in the small intestine (Fig. 55). The first-generation schizonts mature within 60 hours and are found in the epithelial cells of the crypts. The infected cells move into the connective tissue. The adult schizonts measure about 50 by 40 microns. The second generation of larger schizonts (63 by 49 and even 84 microns) forms a multitude of merozoites that are 11.25 microns long and 2.0 microns wide. These appear on the 5th day (120 hours). The sexual stages of development occur on the 6th day and are in the ceca under the nuclei of the epi-

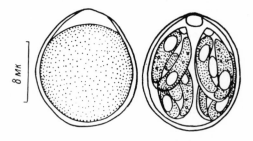

FIG. 55. Oocysts of *E. mivati* (after Edgar and Siebold, 1964).

thelial cells. Development of schizonts may also occur in this location. On the 6th day, the third-generation schizonts form and produce 6 to 16 merozoites (Davies, 1956). The third-generation schizonts are smaller than the second generation. The prepatent period lasts 7 days according to Tyzzer et al. (1932), or 6 days according to Davies (1956) and Edgar (1955); the patent period lasts as long as 12 days. Sometimes after a single infection the oocysts may be passed for an even longer time.

The species has worldwide distribution including the USSR.

*Eimeria praecox* Johnson, 1930. The oocysts are ovoid and have no micropyle. Oocysts measure 23.8 by 20.6 microns (Johnson, 1930). Tyzzer et al. (1932) reported smaller sizes (21.5 by 17.07 microns). There is a polar granule. Sporulation lasts 2 days, or sometimes only 1 day.

Endogenous development occurs in the upper third of the small intestine. The schizonts are under the nuclei of the epithelial cells of the villi. There are two generations of schizonts; the first generation appears within 24 hours, and the second is found on the 2nd day. Each schizont forms from 8 to 24 merozoites. The prepatent period lasts 4 days (84 hours).

The species has worldwide distribution including the USSR.

*Eimeria tenella* (Railliet and Lucet, 1891) Fantham, 1909. The oocysts are ovoid, colorless, and have no micropyle. Dimensions are 14.2–31.2 by 9.5–24.8 microns (mean 22.96 by 19.16 microns) (Becker et al., 1956). According to Scholtyseck (1953), the mean size is 24.5 by 18.2 microns. Gill and Ray (1957) are in opposition to all other investigators in that they observed a residual body in the oocyst. A polar body is present. Sporulation takes 48 hours. The sporocysts measure 8 to 11 microns.

Endogenous development occurs in the ceca but in rare cases it is observed in the lower region of the small intestine and colon. The first-generation schizonts are in the epithelium of the crypts (Fig. 52). The sporozoites pass through the epithelial cells, where they penetrate into macrophages (Pattillo, 1959; Challey and Burns, 1959). They then emerge from the macrophages and again penetrate into epithelial cells at the base of the crypts where schizogony begins. The first-generation merozoites mature on the 2nd or 3rd day. Up to 900 short merozoites about 3 microns long and 1 to 1.5 microns wide form in the mature schizont which measures 24 by 17 microns. On the 4th or 5th day, the second-generation schizonts and merozoites develop. The mature schizonts measure up to 50 microns and sometimes even 87 by 45 microns and contain as many as 200 to 300 merozoites which are 12 to 16 microns long and 2 microns wide. The epithelial cells infected by second-generation schizonts sink into the connective tissue when their development is completed. Scholtyseck (1953) and Greven (1953) reported that schizogony occurs in the fibrocytes. On the 5th or 6th day it is possible to see a few small schizonts 10 to 16 microns in diameter that contain 8 to 16 merozoites which are 4 to 7 microns long. Tyzzer (1929) thought that these were third-generation merozoites, but Gill and Ray (1957) doubted this and suggested that they express sexual differentiation. Scholtyseck denied sexual dimorphism in merozoites. Tyzzer was of the opinion that second-generation merozoites develop partially into gamonts and partially into third-generation schizonts. On the 6th day, gamonts develop in the epithelium. The microgametocytes measure 12.4 by 8.7 microns. The prepatent period is 7 days, and the patent period lasts as long as 10 days. In some cases oocysts have been retained in the ceca for several months (Herrick, 1936).

The species has worldwide distribution and is found in the USSR.

*Wenyonella gallinae* Ray, 1945. The oocysts of this species have been found in chicks in India. The oocysts are ovoid and have a rough wall. There are four sporocysts within the oocyst, each containing four sporozoites. The oocysts are 29.48–33.5 by 19.8–22.78 microns. Sporulation takes 4 to 6 days at 28 C. The prepatent period lasts 7 to 8 days. Oocysts are passed for 3 days.

*Isospora gallinae* Scholtyseck, 1954. The oocysts are ovoid and measure 19–26.5 by 15.5–23 microns (mean 24 by 19.5 microns). They are very similar to oocysts of *I. lacazei,* and experiments conducted by Anpilogova (1965) in infections of chicks with these oocysts demonstrated that infection does take place, and hence, this parasite is not specific for chickens. The oocysts found in chicken feces could have come from sparrows infected with *I. lacazei.*

Found in Europe.

*Cryptosporidium tyzzeri* (Tyzzer, 1912) Levine, 1961. The oocysts are ovoid or spherical, have a single-layered wall, and measure 4–5 by 3 microns. There is no micropyle. The sporozoites are boomerang-shaped and are 5.5 to 6 microns long.

The endogenous stages are on the surface of the epithelial cells, in the striated margin of the tubular tracts of the ceca. The schizonts are 3 to 5 microns in diameter and form up to 8 merozoites which measure 2.5–5 by 0.5–0.7 microns. The microgametocytes are smaller than the schizonts and form 16 microgametes around a residual body. The macrogametes are larger than the microgametocytes.

Found in North America.

## Coccidia of Turkeys

At the present, seven species of *Eimeria,* one species of *Isospora,* and one species of *Cryptosporidium* are known.

*Eimeria adenoeides* Moore and Brown, 1951. Oocysts are usually ellipsoid and slightly asymmetrical. No micropyle is present. Oocysts are 21.5–30.0 by 13.5–19.5 microns (mean 25.7 to 16.25 microns) (Clarkson, 1958). Sporulation takes 24 hours, and 1 to 3 polar granules are present in sporulated oocysts.

Endogenous development occurs in the lower regions of the small intestine, the cecum, and the colon. The first-generation schizonts develop mainly in the narrow portion of the ceca and only a few are in the rectum and small intestine. The second generation is found along the entire cecum and partially in the colon and lower third of the small intestine. All stages are in the epithelium of the villi and the crypts, and second-generation schizonts and gamonts are usually under the nuclei of the epithelial cells and at the base of the crypts (Fig. 56, *A*). First-generation schizonts mature within 60 hours and reach a size of 30 by 18 microns. Up to 700 merozoites develop within each schizont, each measuring 4.5–7 by 1.5 microns. Second-generation schizonts are located above the nuclei of epithelial cells, measure 10 by 10 microns and contain 12 to 24 merozoites which measure 10 by 3 microns. They appear on the 4th day. Gamonts are found on the 5th day. The microgametocytes are the same size as macrogametocytes and have one residual body. The prepatent period lasts 114 to 132 hours; the patent period does not exceed 20 days.

Found in Europe, North America, and India.

*Eimeria gallopavonis* Hawkins, 1952. The oocysts are ellipsoid, measure 22.2–32.7 by 15.2–19.4 microns (average 27.1 by 17.2 microns) and have no micropyle. A polar body is present. Sporulation lasts 24 hours.

Endogenous development occurs in the epithelium of the lower part of the small intestine and in the colon and ceca. On the 3rd day, mature, small, first-generation schizonts with 8 to 12 merozoites are visible. They are found only in the transverse colon and the rectum above the nuclei of the epithelial cells. On the 4th day, the second-generation schizonts appear in the epithelium of the transverse colon, ceca, and rectum. Some schizonts are small, have 10 to 12 merozoites and are found above the nuclei; others reach a size of 25 microns and contain many small merozoites; these develop only in the rectum under the nuclei of the epithelial cells. The gamonts develop mainly in the rectum from the 4th through the 6th day. The prepatent period is 144 hours (6 days).

Found in the USSR, Europe, North America, and India.

*Eimeria innocua* Moore and Brown, 1952. Oocysts are subspherical, have no micropyle and measure 18.6–25.9 by 17.3–24.5 microns (mean 22.4 by 20.9 microns). Sporulation takes 48 hours. No polar granule is formed.

Endogenous development occurs in the small intestine. All stages are in the epithelium of the villi, mainly in the apical part and less often at the base of the crypts. Asexual stages appear on the 4th day after infection. On the 5th day, mature gamonts and oocysts are found. The prepatent period is 114 hours, and the patent period lasts no more than 14 days.

Found in Europe and in North America.

*Eimeria meleagridis* Tyzzer, 1927. Oocysts are ellipsoid, have no micropyle, and are 20.3–30.8 by 15.4–20.6 microns (mean 24.4 by 18.1 microns) (the American strain, according to Tyzzer, 1927). According to Clarkson (1959b), the dimensions of the oocysts of the English strain average 22.4 by 16.25 microns. Sporulation takes 24 hours. There is a polar body.

According to Tyzzer (1927), endogenous development takes place in the ceca and small intestine. However, Clarkson (1959b) has established that first-generation schizonts develop only in the epithelium of the villi in the middle part of the small intestine (Fig. 56, *B*). The second generation is found in the ceca as are the gamonts which are also found in the colon and the lower part of the small intestine. These stages develop in the base of the crypts and in the epithelium on the surface of the ceca. First-generation schizonts mature on the 2nd day after infection, but development may last as long as 5 days. Schizonts are 20 by 15 microns and are under the nuclei of epithelial cells. Fifty to 100 merozoites, which are 7 by 1.5 microns, form within the schizonts. The mature second-generation schizonts appear at the 72nd to 84th hour and measure 9 to 10 microns. They contain 8 to 16 merozoites which are 10 by 2 microns. Hawkins (1952) reported that a third generation is possible, but Clarkson reported that only two generations exist. Within 96 to 108 hours, the mature gamonts appear above the nuclei of the epithelial cells. The microgametocytes are not larger than the macrogametes and contain a residual body. The prepatent period lasts 110 hours.

The species has worldwide distribution including the USSR.

*Eimeria meleagrimitis* Tyzzer, 1929. The oocysts are broadly ovoid, have a smooth wall, and are 16.2–20.5 by 13.2–17.2 microns (mean 18.1 by 15.3 microns) (Tyzzer, 1929). According to Hawkins (1952) the mean is 19.2 by 16.3 microns. Clarkson (1959a) reported the size to be 20.1 by 17.3 microns. During sporulation three polar granules are formed. Sporulation lasts 24 to 48 hours.

Endogenous development (Fig. 56, *C*.) occurs mainly in the upper part of the small intestine. The gamonts, like the schizonts, are concentrated

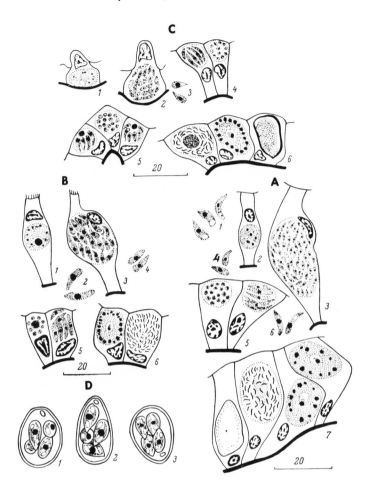

FIG. 56. Endogenous stages of development and oocysts of turkey coccidia (After Clarkson, 1958, 1959a and b). *A, E. adenoeides, 1,* sporozoites; *2,* growing schizont; *3,* adult first-generation schizont at the 60th hour; *4,* first-generation merozoites; *5,* mature second-generation schizont; *6,* second-generation merozoites; *7,* macrogamete, adult microgametocyte and developing oocyst. *B, E. meleagridis, 1,* immature schizont after 36 hours; *2,* first-generation merozoites; *3,* first-generation schizont; *4,* second-generation merozoites; *5,* adult second-generation schizont; *6,* macrogamete and microgametocyte. *C, E. meleagrimitis, 1,* developing first-generation schizont; *2,* adult first-generation schizont; *3,* first-generation merozoites; *4,* mature second-generation schizont; *5,* mature third-generation schizonts; *6,* microgametes, macrogamete, and oocyst. *D,* Oocysts; *1, E. meleagridis; 2, E. adenoeides; 3, E. meleagrimitis.*

in the upper part of the intestine (duodenum and above the small intestine), but they may also be found along the entire length of the small intestine and even in the colon and ceca. First-generation schizonts are in the epithelium of the crypts. The same is observed in the second generation, but in addition they are found in the villi. Third-generation schizonts are in the villi but not in the crypts. The gamonts are also present in the villi, but mainly at the apical ends. First-generation schizonts mature within 48 hours after infection and measure 17 by 13 microns and contain 80 to 100 merozoites which are 4.5 by 1.5 microns. Within 66 hours after infection, second-generation schizonts appear; these measure 8 by 7 microns and contain 8 to 16 merozoites which are 7 microns by 1.5 microns. At the 96th hour, the third generation is found; these are the same size as the second generation. Mature gamonts were found within 114 hours. The microgametocytes have one residual body and are no larger than the macrogametes. The prepatent period is 114 to 118 hours.

The species has worldwide distribution and is found in the USSR.

*Eimeria subrotunda* Moore, Brown and Carter, 1954. The oocysts are subspherical, measure 16.5–26.4 by 14.2–22.4 microns (mean 21.8 by 19.8 microns) and have no micropyle or polar granule. Sporulation takes 48 hours.

Endogenous development occurs in the small intestine, mainly in the duodenum, the ileum, and in the upper part of the transverse colon. All stages are found in the epithelium of the villi, but not in the crypts. The prepatent period is 95 hours, and the patent period lasts 12 to 13 days.

Found in Europe and North America.

*Cryptosporidium meleagridis* Slavin, 1955. The oocysts are ovoid and measure 4.5 by 4 microns. Four sporozoites lie within the oocyst. Sporulation usually takes place within the host.

Endogenous development occurs in the small intestine (primarily in the lower part). All stages are on the surface of the epithelial cells in the striated border. The trophozoites may move into the epithelium. Schizonts measure 4 to 5 microns and contain eight sickle-shaped merozoites which are 5 by 1 micron. The microgametocytes, which are inside epithelial cells, are 4 microns in diameter and form 16 microgametes. Sometimes the microgametocyte lies alongside the macrogamete to undergo a syzygy.

Found in North America.

*Isospora heissini* Svanbaiev, 1955. The oocysts are round and 24.6 to 32.8 microns in diameter. Sporulation takes place in 16 to 20 hours. A polar body is present. The sporocysts measure 14.9 by 10.1 microns. Probably, this is not a valid species, since the oocysts are very similar to *I. lacazei*. Svanbaiev (1955) probably mistook the oocysts of *I. lacazei* from sparrows to be oocysts belonging to turkeys and described them as a new species. The oocysts of *I. lacazei* could easily have fallen into the feces of turkeys accidentally.

Found in the USSR.

*Eimeria dispersa* Tyzzer, 1929. This species was described from the quail, *Colinus virginianus*. Tyzzer succeeded in infecting turkeys with this species. Hawkins (1952) also infected turkeys with this species and studied their development in that host. The oocysts are broadly ovoid and have no micropyles. According to Hawkins (1952), the oocysts are 21.8–31.1 by 17.7–23.9 microns (mean 26 by 21 microns). Sporulation takes 45 hours. The oocysts of *E. dispersa* from turkeys are larger than those of the same species from quail (Fig. 57).

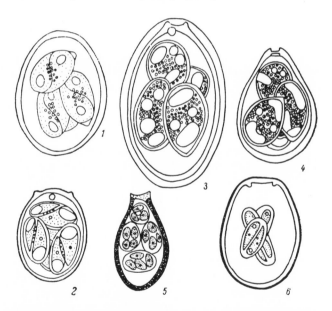

Fig. 57. Oocysts of coccidia of domestic birds. *1, E. dispersa* from turkeys; *2, E. anatis* from ducks; *3, E. nocens; 4, E. anseris* from geese; *5, Wenyonella anatis* from ducks; *6, E. kotlani* from geese (*1* to *4*, and *6*, From Pellérdy, 1965; *5*, After Pande et al. 1965).

Endogenous development takes place in the epithelial cells of the anterior part of the small intestine, mainly in the apical part of the villi. Within about 2 days it is possible to observe first-generation schizonts. Some schizonts are small (6 microns in diameter) and contain 15 merozoites which are 4–6 by 1 micron; others are large (24 by 18 microns) and have 50 merozoites. The second type of schizont is not as common as the first. The second generation matures on the 4th day. The schizonts measure 11 to 13 microns and contain 18 to 23 merozoites which are 5–6 by 1.5–2 microns. Large gamonts are visible on the 5th day. The microgametocytes, which have a residual body, are 20 microns in diameter. The prepatent period is 5 days, but in quail it is only 4 days.

Found in North America.

### Coccidia of Ducks

Coccidia of domestic ducks have been poorly studied, and at the present only three species are known. Two are specific for domestic ducks, and the third is found in wild ducks, but it may also apparently be found in domestic ducks. *Eimeria battakhi* Dubey and Pande (1963) has been described from Indian domestic ducks.

*Eimeria anatis* Scholtyseck, 1955. The oocysts are ovoid and have a distinct micropyle at the narrow end (Fig. 57). The dimensions of the oocysts are 14.4–19.2 by 10.8–15.6 microns (mean 16.8 by 14.1 microns). A polar body and small sporocyst residuum are present. Sporulation lasts 4 days.

Endogenous development takes place in the small intestine. The primary host is the wild duck *Anas platyrhynchas platyrhynchas.*

Found in the USSR, Europe, and North America. Probable in Japan.

*Tyzzeria perniciosa* Allen, 1936. The oocysts are ellipsoid, have no micropyle and measure 10–13.3 by 9–10 microns. The wall is colorless and translucent. Sporulation takes 24 hours. Eight sporozoites, which lie directly within the oocyst, are formed. They are 10 microns long and 3.5 microns wide. A residual body lies between the sporozoites.

Endogenous development takes place in the small intestine. The schizonts are in epithelial cells, but they may penetrate into the connective tissue of the villi. Within 24 hours after infection, mature schizonts measuring 11.6 by 3.3 microns appear; each contains four merozoites. On the 5th day schizonts, which measure 16 by 15 microns, are present.

Allen (1963) was of the opinion that there were three generations of schizonts, the development of which takes place after the formation of the gamonts. The latter appear within 48 hours after infection. The microgametocytes are 8 microns in diameter and are monocentric. The prepatent period is 6 days.

Found in the USSR and North America.

*Wenyonella anatis* Pande, Bhatia, and Srivastava, 1965. The oocysts are ovoid and have a distinct micropyle on the narrow thickened end (Fig. 57). The end of the oocyst with the micropyle is sometimes elongated to form a neck. The oocysts measure 11.5–17 microns by 7–10 microns (mean 14 by 8.5 microns). There is a characteristic puncturing on the surface of the oocyst. The wall is colorless and 0.7 to 1 micron thick. The micropyle is 4.5 to 6 microns wide. Each of the four oval sporocysts contains four sporozoites. The sporocysts are 5.7–7 by 4.3–5 microns (mean 6.3 by 4.7 microns). A polar granule is present. Sporulation takes 48 hours.

Found in India.

## Coccidia of Geese

*Eimeria anseris* Kotlán, 1932. Oocysts are pyriform, have a micropyle at the narrow end and are 16–23 by 13–18 microns. The oocyst wall is thickened around the micropyle (Kotlán, 1932). The oocysts contain a residual body under the micropyle (Klimeš, 1963). The sporocysts measure 10–12 by 7–9 microns (Fig. 57). A residual body is present. Sporulation takes 24 to 48 hours.

Endogenous development occurs in the small intestine. All stages are usually found in the epithelium, but they may penetrate into the connective tissue or even into the muscular layer of the mucosa. The schizonts, which measure 12 to 20 microns, form 15 to 25 merozoites. The microgametocytes are slightly larger than the macrogametes. The prepatent period is 7 days.

World wide distribution includes the USSR.

*Eimeria nocens* Kotlán, 1933. The oocysts are ovoid or ellipsoid and have a micropyle on the narrow end. The external wall is 1.3 microns thick and brown. It is sometimes possible to see a structure which resembles a polar cap above the micropyle. The oocysts are 25–33 by 17–24 microns. According to Hanson, Levine, and Ivens, (1957), the average dimensions are 31 by 21.6 microns. The sporocysts are broadly ellipsoid and 12.1 by 9.1 microns in size (Fig. 57).

Endogenous development occurs in the mucosa of the lower part of the small intestine. All stages are in the epithelium of the villi. The schizonts measure 15 to 30 microns and contain 15 to 35 merozoites. The microgametocytes are larger than the macrogametes and measure 28–36 by 23–31 microns.

Found in the USSR, Europe, and North America.

*Eimeria parvula* Kotlán, 1933. The oocysts are spherical or subspherical, measure 10–15 by 10–14 microns (mean 13 by 10 microns) and have no micropyle.

Endogenous development occurs in the lower part of the small intestine. Klimeš (1963) hypothesized that this species belongs to the genus *Tyzzeria*. According to his data, *T. parvula* forms schizonts in the small intestine with 16 to 20 merozoites. The prepatent period is 5 days.

Found in Europe and North America.

*Eimeria stigmosa* Klimeš, 1963. The oocysts are broadly ovoid, dark brown, and 23 by 16.7 microns. The external wall is thickened at the poles. There is a micropyle and one or two polar granules are present under the micropyle. The sporocysts are egg-shaped and measure 10.7 by 8 microns. Sporulation takes 48 hours.

Endogenous development occurs in the epithelium of the villi in the anterior part of the small intestine. The microgametocytes measure 21 by 17 microns and are the same size as the macrogametes. The prepatent period is 5 days.

Found in Europe.

*E. truncata* (Railliet and Lucet, 1891) Wasielewsky, 1904. The oocysts are ovoid, 20.2 by 13–16 microns and have one thickened pole. The micropyle measures 1.5 microns. The sporocysts are 7.8 by 5.5 microns and have a residual body. The residual body is not present in all oocysts, but it may be there on rare occasions (Levine, 1961a). Sporulation lasts 48 hours, and perhaps as long as 5 days.

Endogenous development occurs in the epithelium of the renal tubules. Schizonts are 13 microns and form 20 to 30 merozoites which are 4–6 by 1.4–2 microns. The microgametocytes are 7 to 13 microns in diameter. The prepatent period is 5 to 6 days (Kotlán, 1932).

Distribution is probably worldwide and includes the USSR.

*E. kotlani* Gräfner and Graubmann, 1964. The oocysts are broadly oval or ovoid and have a micropyle surrounded by two lip-shaped thickenings of the internal wall. Oocysts are 29–33 by 23–25 microns (Fig. 57). The oocyst wall is 2 microns thick. Sporulation lasts 4 days. A sporocyst residuum is present.

Endogenous development takes place in the large intestine and the tubular parts of the ceca. Schizonts and gamonts begin development in the epithelium, but these stages then move in to the subepithelium as far as the muscular layer. Mature schizonts are 14.6–20.8 microns, and the microgametocytes are 16.6–31.2 microns.

Found in East and West Germany.

*Tyzzeria anseris* Nieschulz, 1947. Oocysts are ellipsoidal and 12–16 by 10–12.5 microns (mean 14 by 11.5 microns). Eight sporozoites lie around the residual body of the oocyst.

Found in the USSR, Europe, and North America.

Oocysts of other species of coccidia specific for wild geese may be found in domestic geese (cf. Pellérdy, 1965).

# BIBLIOGRAPHY

# BIBLIOGRAPHY

Aikawa, M., P. Hepler, C. Huff, and H. Sprinz. 1966. The feeding mechanism of avian malarial parasites. J. Cell Biol. 28:355–375.

Albanese, A. A., and H. Smetana. 1937. Studies on the effect of X-rays on the pathogenicity of *Eimeria tenella*. Amer. J. Hyg. 26:27–29.

Allen, E. A. 1934. *Eimeria angusta* sp. nov. and *Eimeria bonasae* sp. nov. from grouse, with a key to the species of *Eimeria* in birds. Amer. Microscop. Soc., Trans. 53:1–5.

Allen, E. A. 1936. *Tyzzeria perniciosa* gen. et sp. nov., a coccidium from the small intestine of the Pekin duck, *Anas domesticus* L. Arch. Protistenk. 87:262–267.

Andrews, J. M. 1926. A factor in host-parasite specificity of coccidiosis. Anat. Rec. 34: 154.

Andrews, J. M. 1930. Excystation of coccidial oocyst in vivo. Science n.s. (1828), 71:37.

Andrews, J. M. 1933. The control of poultry coccidiosis by the chemical treatment of litter. Amer. J. Hyg. 17:466–490.

Anpilogova, N. V. 1965. On the finding of oocysts of coccidia of the genus *Isospora* in domestic birds. [in Russian]. Dokl. AN TadzhSSR. 8:44–46.

Avery, J. L. 1942. The effect of moderately low temperatures on the sporulation of oocysts of two species of swine coccidia. J. Parasitol. 28(Suppl.):28.

Baker, D. W. 1938. Observations on *Eimeria bukidnonensis* in New York State cattle. J. Parasitol. 24 (Suppl.):15–16.

Balozet, L. 1932. Les coccidies du mouton et de la chevre. Cycle evolutif de *Eimeria ninakohl-yakimovi*. Arch. Inst. Pasteur Tunis 21:88–118.

Bano, L. 1959. A cytological study of the early oocysts of seven species of Plasmodium and the occurrence of post-zygotic meiosis. Parasitology 49:559–585.

Beach, J. R., and D. E. Davis. 1925. Coccidiosis of chickens. Univ. Calif. Agr. Exp. Sta. Circ. 300. 16 p.

Becker, E. R. 1934. Coccidia and of coccidiosis of domesticated, game and laboratory animals and of man. Monogr. 2, Div. Ind. Sci., Iowa State Coll., Ames.

Becker, E. R. 1952. Protozoa, p. 947–976. *In* Biester and Schwarte, Diseases of poultry. 2nd Ed., Iowa State Coll. Press, Ames.

Becker, E. R., and H. B. Crouch. 1931. Some effects of temperature upon development of the oocysts of coccidia. Proc. Soc. Exp. Biol. Med. 28:529–530.

Becker, E. R., and P. R. Hall. 1933a. Cross-immunity and correlation of oocyst production during immunization between *Eimeria miyairii* and *Eimeria separata* in the rat. Amer. J. Hyg. 18:220–223.

Becker, E. R., and P. R. Hall. 1933b. Quantitative character of coccidian infection in recipients of blood from immunized animals. Proc. Soc. Exp. Biol. Med. 30:789–790.

Becker, E. R., P. R. Hall, and A. Hager. 1932. Quantitative, biometric and host-parasite studies on *Eimeria miyairii* and *Eimeria separata* in rats. Iowa State Coll. J. Sci. 6: 299–316.

Becker, E. R., R. J. Jessen, W. H. Pattillo, and W. M. Van Doorninck. 1956. A biometrical study of the oocyst of *Eimeria necatrix*, a parasite of the common fowl. J. Protozool. 3:126–131.

Becker, E. R., and N. F. Morehouse. 1937. Modifying the quantitative character of coccidian infection through a dietary factor. J. Parasitol. 23:153–162.

Becker, E. R., W. H. Pattillo, J. N. Farmer, and W. M. Van Doorninck. 1955. Size of the oocyst of *Eimeria brunetti* and *Eimeria necatrix.* J. Parasitol. 41(Suppl.):18.

Becker, E. R., W. J. Zimmermann, and W. H. Pattillo. 1955. A biometrical study of the oocyst of *Eimeria brunetti,* a parasite of the common fowl. J. Protozool. 2:145–150.

Becker, E. R., W. J. Zimmermann, W. H. Pattillo, and J. N. Farmer. 1956. Measurements of the unsporulated oocysts of *Eimeria acervulina, E. maxima, E. tenella* and *E. mitis;* coccidian parasites of the common fowl. Iowa State Coll. J. Sci. 31:79–84.

Bělăr, K. 1926a. Der Formwechsel der Protistenkerne. Eine vergleichend-morphologische Studie. Ergebn. Fortschr. Zool. 6:235–654.

Bělăr, K. 1926b. Zur Cytologie von *Aggregata eberthi.* Arch. Protistenk. 53:312–325.

Berghe, L. van den, 1946. A cytochemical study of the "volutin granules" in protozoa. J. Parasitol. 32:465–466.

Beyer, T. V. 1960. Cytological research on various stages in the life cycle of rabbit coccidia. Phosphomonoesterases in *Eimeria magna.* [in Russian]. Voprosy Tsitologii i Protistologgi, Izd. AN SSSR, M.-L.:277–284.

Beyer, T. V. 1962. On the distribution of succino-dehydrogenase in the life cycle of *Eimeria intestinalis* [in Russian]. Tsitologiia 4:232–237.

Beyer, T. V. 1963a. Cytochemical investigation of thiol compounds in various stages of development of *Eimeria intestinalis* [in Russian]. Tsitologiia 5:61–68.

Beyer, T. V. 1963b. Change in succino-dehydrogenase activity at various stages of development of rabbit coccidia under the effects of furacillin [in Russian]. Morfologiya i Fiziologiya Prosteyshikh, Izd. AN SSSR, M.-L.:68–79.

Beyer, T. V., and L. P. Ovchinnikova. 1964. A cytophotometrical investigation of the RNA content in the course of macrogametogenesis in two rabbit intestinal coccidia *Eimeria magna* and *E. intestinalis.* Acta Protozool. 2:329–337.

Beyer, T. V., and L. P. Ovchinnikova. 1966. A cytophotometrical investigation of the cytoplasmic RNA content in the course of oocyst formation in the intestinal rabbit coccidia *Eimeria intestinalis* Cheissin, 1948. Acta Protozool. 4:75–80.

Bhatia, B. L. 1938. Protozoa: Sporozoa. *In* The fauna of British India, Taylor and Francis, Ltd., London. 497 p.

Biester, H. E., and C. Murray. 1934. Studies in infectious enteritis of swine. VIII. *Isospora suis* n. sp. in swine. J. Amer. Vet. Med. Assoc. 85:207–219.

Boch, J., E. Pezenburg, and V. Rosenfeld. 1961. Ein Beitrag zur Kenntnis der Kokzidien der Schweine. Berlin Munchen Tierärztl. Wochenshr. 74:449–451.

Boch, J. and E. Wiesenhütter. 1963. Über einzelne Stadien der endogenen Entwicklung des Schweinekokzids *Eimeria debliecki* Douwes, 1921. Berlin Munchen Tierärztl. Wochenschr. 76:236–238.

Boles, J. I., and E. R. Becker. 1954. The development of *Eimeria brunetti* Levine in the digestive tract of chickens. Iowa State Coll. J. Sci. 29:1–26.

Bondois, C. M. 1936. Contribution a l'etude de la coccidiose intestinale du pigeon-voyageur. Trav. Lab. Zool. Parasitol. Fac. Med. Lille:29–36.

Borchert, A. 1958. Lehrbuch der Parasitologie fur Tierarzte. Herzel Verlag, Leipzig. 454 p.

Boughton, D. C. 1929. A note on coccidiosis in sparrows and poultry. Poultry Sci. 8:184–188.

Boughton, D. C. 1930. The value of measurements in the study of a protozoan parasite *Isospora lacazie.* Amer. J. Hyg. 11:212–226.

Boughton, D. C. 1933. Diurnal gametic periodicity in avian *Isospora.* Amer. J. Hyg. 18:161–184.

Boughton, D. C. 1934. Periodicity in the oocyst-production of the pigeon *Eimeria*. J. Parasitol. 20:329.

Boughton, D. C. 1937a. Studies on oocyst production in avian coccidiosis. II. Chronic isosporan infections in the sparrow. Amer. J. Hyg. 25:203–211.

Boughton, D. C. 1937b. Studies on oocyst production in avian coccidiosis. III. Periodicity in the oocyst production of eimerian infections in the pigeon. J. Parasitol. 23:291–293.

Boughton, D. C. 1937c. Notes on avian coccidiosis. The Auk. 54:500–509.

Boughton, D. C. 1939. The sporulation of oocysts in various types of litter. Bull. Univ. Georgia 39:9–14.

Boughton, D. C., F. O. Atchley, and L. C. Eskridge. 1935. Experimental modification of the diurnal oocyst-production of the sparrow coccidium. J. Exp. Zool. 70:55–74.

Bovee, E. C. 1962. *Isospora crotali* in the Florida diamond-back rattlesnake. J. Protozool. 9:459–466.

Bovee, E. C. 1965. Swimming movements of *Eimeria* sp. merozoites. Progress in Protozoology. 2nd Int. Conf. Protozool., London, Abstr. 1952.

Bovee, E. C., and S. R. Telford, Jr. 1965a. *Eimeria sceloporis* and *Eimeria molochis* spp. n. from lizards. J. Parasitol. 51:85–94.

Bovee, E. C., and S. R. Telford, Jr. 1965b. *Eimeria noctisauris* sp. n., a coccidian from the lizard, *Klauberina riversiana*. J. Parasitol. 51:325–330.

Brackett, S., and A. Bliznick. 1949. The effect of small doses of drugs on oocyst production of infections with *Eimeria tenella*. Ann. N.Y. Acad. Sci. 52:595–610.

Brackett, S., and A. Bliznick. 1952. The reproductive potential of five species of coccidia of the chicken as demonstrated by oocyst production. J. Parasitol. 38:133–139.

Braden, A. 1955. The reactions of isolated mucopolysaccharides to several histochemical tests. Stain Techn. 30:19–26.

Bray, R. S. 1957. Studies on the exo-erythrocytic cycle in the genus *Plasmodium*. London Sch. Hyg. Trop. Med., Memoir 12. 192 p.

Brotherston, J. G. 1948. The effect of relative dryness on the oocysts of *Eimeria tenella* and *Eimeria bovis*. Roy. Soc. Trop. Med. Hyg., Trans., 42:10–11.

Brumpt, E. 1949. Précis de Parasitologie. Masson et Cie, Paris. 2138 p.

Canning, E. U. 1962. Sexual differentiation of merozoites of *Barrouxia schneideri* (Butschli). Nature 195:720–721.

Carvalho, J. C. 1942. *Eimeria neoleporis* n. sp. occurring naturally in the cottontail and transmissible to the tame rabbit. Iowa State Coll. J. Sci. 16:409–410.

Carvalho, J. C. 1943. The coccidia of wild rabbits of Iowa. I. Taxonomy and host-specificity. Iowa State Coll. J. Sci. 18:103–135.

Carvalho, J. C. 1944. The coccidia of wild rabbits of Iowa. II. Experimental studies with *Eimeria neoleporis* Carvalho, 1942. Iowa State Coll. J. Sci. 18:177–189.

Chakravarty, M., and I. B. Kar. 1946. Effect of temperature on the sporulation and mortality of coccidian oocysts. Proc. Nat. Inst. Sci. India 12:1–6.

Challey, J. R., and W. H. Burns. 1959. The invasion of the cecal mucosa by *Eimeria tenella* sporozoites and their transport by macrophages. J. Protozool. 6:238–241.

Chang, K. 1937. Effects of temperature on the oocysts of various species of *Eimeria* (Coccidia, Protozoa). Amer. J. Hyg. 26:337–351.

Chatton, E. 1910. Le kyste de Gilruth dans la muqueuse stomacale des ovidés. Arch. Zool. Exp. Gen. 5:114–124.

Chatton, E. and F. Villeneuve. 1936. Le cycle évolutif de l'*Eleutheroschizon duboscqui* Brazil. Preuve expérimentale de l'absence de schizogonie chez cette forme et chez la *Siedieckia caulleryi* ch. et villen. C. R. Acad. Sci. (Paris) 203:833–836.

Christensen, J. F. 1939. Sporulation and viability of oocysts of *Eimeria arloingi* from the domestic sheep. J. Agr. Res. 59:527–534.

Christensen, J. F. 1941. The oocysts of coccidia from domestic cattle in Alabama (U.S.A.), with descriptions of two new species. J. Parasitol. 27:203–220.

Clarkson, M. J. 1958. Life history and pathogenicity of *Eimeria adenoeides* Morre and Brown, 1951, in the turkey poult. Parasitology 48:70–88.

Clarkson, M. J. 1959a. The life history and pathogenicity of *Eimeria meleagrimitis* Tyzzer, 1929, in the turkey poult. Parasitology 49:70–82.

Clarkson, M. J. 1959b. The life history and pathogenicity of *Eimeria meleagridis* Tyzzer, 1927, in the turkey poult. Parasitology 49:519–528.

Clarkson, M. J. 1960. The coccidia of the turkey. Ann. Trop. Med. Parasitol. 54:253–257.

Clarkson, M. J. 1963. Coccidia and coccidiosis in the turkey. 1st Int. Conf. Protozool., 1961, Prague, Proc.:452–455.

Clarkson, M. J., and M. A. Gentles. 1958. Coccidiosis in turkeys. Vet. Rec. 70:211–214.

Cordero del Campillo, M. 1959. Estudios sobre *Eimeria falciformis* (Eimer, 1870) parasito del raton. Rev. Iber. Parasitol. 19:351–368.

Curasson, G. 1943. Traité de protozoologie vétérinaire et comparée. I. Typanosomes. II. Spirochétes, flagellés, infusoires, rhizopodes. III. Sporozoaires. 445, 330, 493 pp. Paris.

Dasgupta, B. 1959. The feulgen reaction in the different stages of the life-cycles of certain Sporozoa. Quart. J. Microscop. Sci. 100:241–255.

Dasgupta, B. 1960. Polysaccharides in the different stages of the life-cycles of certain Sporozoa. Parasitology 50:509–514.

Dasgupta, B. 1961a. Alkaline phosphatase reaction in the different stages of the life-cycles of certain Sporozoa. Vestnik Ceskosl. Zool. Spolec., Acta Soc. Zool., Bohemo-Slov. 25: 16–21.

Dasgupta, B. 1961b. The basophilia in the different stages of the life-cycles of certain Sporozoa. Vestnik Ceskosl. Zool. Spolec., Acta Soc. Zool., Bohemo-Slov. 25:203–214.

Davies, S. F. M. 1956. Intestinal coccidiosis in chickens caused by *Eimeria necatrix*. Vet. Rec. 68:853–857.

Davies, S. F. M., and L. P. Joyner. 1962. Infection of the fowl by the parenteral inoculation of oocysts of *Eimeria*. Nature 194:996–997.

Davies, S. F. M., L. P. Joyner, and S. B. Kendall. 1963. Coccidiosis. Oliver & Boyd, Ltd., Edinburgh. 264 p.

Davis, L. R., D. C. Boughton, and G. W. Bowman. 1955. Biology and pathogenicity of *Eimeria alabamensis* (Christensen, 1941), an intranuclear coccidium of cattle. Amer. J. Vet. Res. 16:274–281.

Davis, L. R., and G. W. Bowman. 1957. The endogenous development of *Eimeria zurnii*, a pathogenic coccidium of cattle. Amer. J. Vet. Res. 18:569–574.

Davis, L. R., and G. W. Bowman. 1962. Schizonts and microgametocytes of *Eimeria auburnensis* Christensen and Porter, 1939, in calves. J. Protozool. 9:424–427.

Davis, L. R., and G. W. Bowman. 1964. Observations on the life cycle of *Eimeria bukidnonensis* Tubanqui, 1931, a coccidium of cattle. J. Protozool. 11(Suppl.):17.

Davis, L. R., and G. W. Bowman. 1965. The life history of *Eimeria intricata* Spiegl, 1925 in domestic sheep. Progress in Protozool. 2nd Int. Conf. Protozool., London, Abstr. 160.

Deiana, S., and G. Delitala. 1953. La coccidiosi dei piccoli ruminanti. Nota I: Rilievi morfologici e biologici sui coccidi repertati in alcuni caprini della Sardegna (*Eimeria arloingi*, Marotel 1950) Atti. Soc. Ital. Sci. Vet. 7:616–618.

Delaplane, J. R., and H. O. Stuart. 1935. The survival of avian coccidia in soil. Poultry Sci. 14:67–69.

Deom, J., and J. Mortelmans. 1954. Observations sur la coccidiose du pore a *Eimeria debliecki* au Congo Belge. Ann. Soc. Belg. Med. Trop. 34:43–46.

Dobell, C. 1925. The life history and chromosome cycle of *Aggregata eberthi* [Protozoa: Sporozoa: Coccidia]. Parasitology 17: 1–136.

Dobell, C., and A. P. Jameson. 1915. The chromosome cycle in coccidia and gregarines. Proc. Roy. Soc. London 89:83–94.

Doflein, F., and E. Reichenow. 1953. Lehrbuch der Protozoenkunde. 6th Edition. VEB Gustav Fischer Verlag, Jena. 1213 p.

Dogel', V. A. 1948. Parasitic protozoa of the fishes in the Gulf of Peter the Great [in Russian]. Izv. Vsesoyuzn. N.-Issl. Inst. Ozern. Rechn. Rybn. Khoz. 27: 17–66.

Dogel', V. A. 1949. The phenomenon of "species conjugation" in parasites and the evolutionary significance of this phenomenon. [in Russian]. Izv. AN KazSSR, Ser. Parazitol. 74:3–15.

Dogel', V. A., and A. KH. Akhmerov. 1946. Parasitofauna of the fishes of the Amur and its zoogeographical significance [in Russian]. Tr. Yubileynoy Sess. LGU, Sekts. Biol.: 171–178.

Dogel', V. A., Yu. I. Polyansky, and Ye. M. Kheysin. 1962. General protozoology [in Russian]. Izd. AN SSSR, M.-L.: 1–591.

Doran, D. J. 1966a. The migration of *Eimeria acervulina* sporozoites to the duodenal glands of Lieberkühn. J. Protozool. 13:27–33.

Doran, D. J. 1966b. Location and time of penetration of duodenal epithelial cells by *Eimeria acervulina* sporozoites. Proc. Helminthol. Soc. Wash. 33: 43–46.

Doran, D. J., and M. M. Farr. 1961. In vitro excystation of *Eimeria acervulina*. J. Parasitol. 47(Suppl.):45.

Doran, D. J., and M. M. Farr. 1962a. Excystation of the poultry coccidium, *Eimeria acervulina*. J. Protozool. 9:154–161.

Doran, D. J., and M. M. Farr. 1962b. *Eimeria acervulina* infections in 1, 2 and 3 day-old chicks. J. Parasitol. 48(Suppl. 2):33.

Doran, D. J., and T. L. Jahn. 1952. Preliminary observations on *Eimeria mohavensis* n. sp. from the kangaroo rat *Dipodomys panamintinus mohavensis* (Grinnell). Amer. Micr. Soc., Trans. 71:93–101.

Doran, D. J., T. L. Jahn, and R. Rinaldi. 1962. Excystation and locomotion of *Eimeria acervulina* sporozoites. J. Parasitol. 48(Suppl.):32–33.

Duncan, S. 1957. Some aspects of the biology of the pigeon coccidium, *Eimeria labbeana* Pinto, 1928. Diss. Abstr. 17:1629.

Duncan, S. 1959a. The effects of some chemical and physical agents on the oocysts of the pigeon coccidium, *Eimeria labbeana* (Pinto, 1928). J. Parasitol. 45:193–197.

Duncan, S. 1959b. The size of the oocysts of *Eimeria labbeana*. J. Parasitol. 45:191–192.

Edgar, S. A. 1954. Effect of temperature on the sporulation of oocysts of the protozoan, *Eimeria tenella*. Amer. Microscop. Soc., Trans. 73:237–242.

Edgar, S. A. 1955. Sporulation of oocysts at specific temperatures and notes on the prepatent period of several species of avian coccidia. J. Parasitol. 41: 214–216.

Edgar, S. A., C. A. Herrick, and L. A. Fraser. 1944. Glycogen in the life cycle of the coccidium, *Eimeria tenella*. Amer. Microscop. Soc., Trans. 63:199–202.

Edgar, S. A., and C. T. Seibold. 1964. A new coccidium of chickens, *Eimeria mivati* sp. n. (Protozoa: Eimeriidae) with details of its life history. J. Parasitol. 50:193–204.

Ellis, C. C. 1938. Studies of the variability of the oocysts of *Eimeria tenella*, with particular reference to the conditions of incubation. Cornell Vet. 28:267–272.

Enigk, K. 1934. Zur kenntnis des *Globidium cameli* und der *Eimeria cameli*. Arch. Protistenk. 83:371–380.

Farr, M. M. 1953. Three new species of coccidia from the Canada goose, *Branta canadensis*. J. Wash. Acad. Sci. 43:336–340.

Farr, M. M. 1964. Life cycle of *Eimeria gallopavonis* (Hawkins) in the turkey. J. Parasitol. 50(Suppl.):52.

Farr, M. M., and D. J. Doran, 1961a. Comparative studies on in vitro excystation of some avian coccidia. J. Protozool. 8(Suppl.):10.

Farr, M. M., and D. J. Doran. 1961b. In vivo excystation of *Eimeria acervulina*. J. Parasitol. 47(Suppl.):45.

Farr, M. M., and D. J. Doran. 1962a. Observations on excystation of poultry coccidia. Virginia J. Sci. (n.s.) 13:211–212.

Farr, M. M., and D. J. Doran. 1962b. Comparative excystation of four species of poultry coccidia. J. Protozool. 9:403–407.

Farr, M. M., and E. E. Wehr. 1949. Survival of *Eimeria acervulina, Eimeria tenella,* and *Eimeria maxima* oocysts on soil under various field conditions. Ann. N.Y. Acad. Sci. 52:468–472.

Fawcett, D. W., and K. R. Porter. 1954. A study of the fine structure of ciliated epithelia. J. Morph. 94:221–282.

Fiebiger, J. 1913. Studien über die Schwimmblasencoccidien der Gadusarten (*Eimeria gadi* n. sp.). Arch. Protistenk. 31:95–137.

Fish, F. F. 1931a. The effect of physical and chemical agents on the oocysts of *Eimeria tenella*. Science 73:292–293.

Fish, F. F. 1931b. Quantitative and statistical analysis of infections with *Eimeria tenella* in the chicken. Amer. J. Hyg. 14:560–576.

Fitzgerald, P. R. 1962. The results of intraperitoneal or intramuscular injections of sporulated or unsporulated oocysts of *Eimeria bovis* in calves. J. Protozool. 9(Suppl.):21–22.

Fitzgerald, P. R. 1965. The results of parenteral injections of sporulated or unsporulated oocysts of *Eimeria bovis* in calves. J. Protozool. 12:214–221.

Galli-Valerio, B. 1918. Coccidia in the intestines—red dysentery of cattle—observations on *Eimeria zurni-rivolta*. Vet. J. 74:219–223.

Garnham, P. C. C., J. R. Baker, and R. G. Bird. 1962. The fine structure of *Lankesterella garnhami*. J. Protozool. 9:107–114.

Garnham, P. C. C., R. G. Bird, and J. R. Baker. 1963. Electron microscope studies of motile stages of malaria parasites. IV. The fine structure of the sporozoites of four species of *Plasmodium*. Roy. Soc. Trop. Med. Hyg., Trans. 57:27–31.

Gill, B. S. 1954. Speciation and viability of poultry coccidia in 120 faecal samples preserved in 2.5 per cent potassium dichromate solution. Indian J. Vet. Sci. Anim. Husb. 24:245–247.

Gill, B. S., and H. N. Ray. 1954a. Phosphatases and their significance in *Eimeria tenella* Railliet et Lucet, 1891. Indian J. Vet. Sci. Anim. Husb. 24:239–244.

Gill, B. S., and H. N. Ray. 1954b. Glycogen and its probable significance in *Eimeria tenella* Railliet et Lucet, 1891. Indian J. Vet. Sci. Anim. Husb. 24:223–228.

Gill, B. S., and H. N. Ray. 1954c. On the occurrence of mucopolysaccharides in *Eimeria tenella* Railliet et Lucet, 1891. Indian J. Vet. Sci. Anim. Husb. 24:229–237.

Gill, B. S., and H. N. Ray. 1957. Life-cycle and cytology of *Eimeria tenella* Railliet et Lucet, 1891 (Protozoa: Sporozoa), with notes on symptomatology and pathology of the infection. Zool. Soc. Calc., Proc., Mookerjee Memor. Vol.: 357–383.

Gill, G. S., and H. N. Ray. 1960. The coccidia of domestic rabbit and the common field hare of India. Zool. Soc. Calc., Proc. 13:129–142.

Giovannola, A. 1934. Die Glykogenreaktionen nach Best und nach Bauer, in ihrer Anwendung auf Protozoen. Arch. Protistenk. 83:270–274.

Goodrich, H. P. 1944. Coccidian oocysts. Parasitology 36:72–79.

Gräfner, G., and H.-D. Graubmann. 1964. *Eimeria kotlani* n. sp. eine neue pathogene Kokzidienart bei Gänsen. Monatsh. Vet. Med. 19:819–821.

Grassé, P. P. 1953. Sous-embranchement des sporozoaires. *In* Traité de Zoologie, 1, Fasc. 2:545–906.

Greiner, J. 1921. Cytologische Untersuchung der Gametenbildung und Befruchtung von *Adelea ovata (A. schneider)*. Zool. Jahrb., Jena, Abt. Anat. 42:327–362.

Grell, K. G. 1953. Entwicklung und Geschlechtsbestimmung von *Eucoccidium dinophili*. Arch. Protistenk. 99:156–186.

Grell, K. G. 1956. Protozoologie. Springer-Verlag OHG, N Berlin. 1–284.

Grell, K. G. 1960. Reziproke Infektion mit *Eucoccidien* aus Verschiedenen Wirten. Naturwissenschaften 47:47–48.

Greven, U. 1953. Zur Pathologie der Geflügelcoccidiose. Arch. Protistenk. 98:342–414.

Hadley, P. B., and E. E. Amison. 1911. *Eimeria avium:* A morphological study. Arch. Protistenk. 23:7–50.

Hagan, K. W., Jr., 1958. The effects of age and temperature on the survival of oocysts of *Eimeria stiedae*. Amer. J. Vet. Res. 19:1013–1014.

Hall, P. R. 1933. Notes on the sporulation time, prepatent period, patent period, and size of oocysts in infections of *Isospora lacazei* Labbé. Iowa Acad. Sci., Proc. 40:221–225.

Hall, P. R. 1934. The relation of the size of the infective dose to number of oocysts eliminated, duration of infection, and immunity in *Eimeria miyairii* Ohira infections in the white rat. Iowa State Coll. J. Sci. 9:115–124.

Hall, R. P. 1953. Protozoology. Prentice-Hall Inc., Eaglewood Cliffs, N.J., 692 p.

Hammond, D. M. 1964. Coccidiosis of cattle, some unsolved problems. 30th Faculty Honor Lecture. The Faculty Assoc., Utah State Univ., Logan. 38 p.

Hammond, D. M., F. L. Andersen, and M. L. Miner. 1963. The occurrence of a second asexual generation in the life cycle of *Eimeria bovis* in calves. J. Parasitol. 49:428–434.

Hammond, D. M., G. W. Bowman, L. R. Davis, and B. T. Simms. 1946. The endogenous phase of the life cycle of *Eimeria bovis*. J. Parsitol. 32:409–427.

Hammond, D. M., W. N. Clark, and M. L. Miner. 1961. Endogenous phase of the life cycle of *Eimeria auburnensis* in calves. J. Parasitol. 47:591–596.

Hammond, D. M., and J. V. Ernst. 1964. Cytological observations on *Eimeria bovis* merozoites. J. Protozool. 11:17–18.

Hammond, D. M., T. W. McCowin, and J. L. Shupe. 1954. Effect of site of inoculation and of treatment with sulfathalidine-arsenic on experimental infection with *Eimeria bovis* in calves. Utah Acad. Sci., Proc. 31:161–162.

Hammond, D. M., F. Sayin, and M. L. Miner. 1963. Über den Entwicklungszyklus und die Pathogenität von *Eimeria ellipsoidalis* Becker und Frye, 1929, in Kalbern. Berlin Munchen Tierärztl. Wochenschr. 76:331–332.

Hanson, H. C., N. D. Levine, and V. Ivens. 1957. Coccidia (Protozoa: Eimeriidae) of North American wild geese and swans. Canad. J. Zool. 35:715–733.

Hardcastle, A. B. 1944. *Eimeria brevoortiana*, a new sporozoan parasite from menhaden *(Brevoortia tyrannus)* with observations on its life history. J. Parasitol. 30:60–68.

Hauschka, T. S. 1943. Life history and chromosome cycle of the coccidian *Adelina deronis*. J. Morphol. 73:529–581.

Hawkins, P. A. 1952. Coccidiosis in turkeys. Techn. Bull. Mich. Agr. Exp. Sta. 226:1–87.

Hemmert-Halswick, A. 1943. Infection mit *Globidium leuckarti* beim Pferd. Z. Veterinark 55:192–199.

Henneré, E. 1963. Cycle biologique de *Coelotropha vivieri*, n. gen. n. sp., coccidie parasite de *Notomastus latericeus* Sars. C. R. Acad. Sci. (Paris) 256:3204–3206.

Henneré, E., and E. Vivier. 1962. Phenomènes d'exuviation chez une coccidie. *Eucoccidium durchoni* Vivier, parasite de *Nereis diversicolor.* C. R. Acad. Sci. (Paris) 255: 564–566.

Henry, A. C. L., and G. Masson. 1932. Considération sur le genre *Globidium; Globidium cameli* n. sp. parasite du dromadaire. Ann. Parasitol. 10:385–401.

Henry, D. P. 1931. A study of the species of *Eimeria* occurring in swine. Univ. Calif. Publ. Zool. 36:115–126.

Henry, D. P. 1932a. Coccidiosis of the guinea pig. Univ. Calif. Publ. Zool. 37:211–268.

Henry, D. P. 1932b. The oocyst wall in the genus *Eimeria.* Univ. Calif. Publ. Zool. 37: 269–278.

Herrick, C. A. 1935. Resistance of the oocysts of *Eimeria tenella* to incubator conditions. Poultry Sci. 14:246–252.

Herrick, C. A. 1936. Organ specificity of the parasite *Eimeria tenella.* J. Parasitol. 22: 226–227.

Herrick, C. A., G. L. Ott, and C. E. Holmes. 1936. Age as a factor in the development of resistance of the chicken to the effects of the protozoan parasite, *Eimeria tenella.* J. Parasitol. 22:264–272.

Hitchcock, D. J. 1955. The life cycle of *Isospora felis* in the kitten. J. Parasitol. 41: 383–397.

Hoare, C. A. 1927. On the coccidia of the ferret. Ann. Trop. Med. Parasitol. 21:313–320.

Hoare, C. A. 1933. Studies on some new ophidian and avian coccidia from Uganda, with a revision of the classification of the Eimeriidea. Parasitology 25:359–388.

Hoare, C. A. 1935. The endogenous development of the coccidia of the ferret, and the histopathological reaction of the infected intestinal villi. Ann. Trop. Med. Parasitol. 29: 111–122.

Hoare, C. A. 1956. Classification of coccidia Eimeriidae in a "Periodic System" of homologous genera. Rev. Bras. Malar. 8:198–202.

Holz, J. 1954. Elektronenmikroskopische Untersuchungen über die Wirksamkeit verschidener Waschmittel auf die Oocysten einiger Coccidienarten. Tierarztl. Umsch. 9:415–419.

Horton-Smith, C. 1947. Coccidiosis—some factors influencing its epidemiology. Vet. Rec. 59:645–646.

Horton-Smith, C. 1954. Parasites and deep litter. Canad. Poultry Rev. 78:46–48.

Horton-Smith, C., and P. L. Long. 1963. Coccidia and coccidiosis in the domestic fowl and turkey. *In* Advances Parasitol., 1, Acad. Press. p. 67–107.

Horton-Smith, C., and P. L. Long. 1965. The development of *Eimeria necatrix* Johnson, 1930 and *Eimeria brunetti* Levine, 1942 in the caeca of the domestic fowl *(Gallus domesticus).* Parasitology 55:401–407.

Horton-Smith, C., E. L. Taylor, and E. E. Turtle. 1940. Ammonia fumigation for coccidial disinfection. Vet. Rec. 52:829–832.

Hosoda, S. 1928. Experimentelle Studien über die Entwickelung des *Eimeria avium.* Fuknoka-Ikwadaigaku Zasshi 21:777–848.

Ikeda, I. 1914. Studies on some sporozoan parasites of sipunculoids. II. *Dobellia binucleata* n. g. n. sp.; a new coccidian from the gut of *Petalostoma minutum.* Arch. Protistenk. 33:205–246.

Ikeda, M. 1955a. Factors necessary for *Eimeria tenella* infection of the chicken. I. Influence of the digestive juices on infection. Jap. J. Vet. Sci. 17:197–200.

Ikeda, M. 1955b. Factors necessary for *E. tenella* infection of the chicken. I. The influence of the digestive juice on infections. Rec. Researches Fac. Agr. Univ. Tokyo 4:98.

Ikeda, M. 1956a. Factors necessary for *E. tenella* infection of the chicken. III. Influence of the upper alimentary canal on infection. Jap. J. Vet. Sci. 18:25–30.

Ikeda, M. 1956b. IV. Investigations of the site of the action of the pancreatic juice on infection. Jap. J. Vet. Sci. 18:45–52.

Ikeda, M. 1957. Factors necessary for *E. tenella* infection of the chicken. V. Organ specificity of *E. tenella* infection. Jap. J. Vet. Sci. 19:105–113.

Ikeda, M. 1960a. Factors necessary for *Eimeria tenella* infection of the chicken. VI. Excystation of oocyst in vitro. Jap. J. Vet. Sci. 22:27–41.

Ikeda, M. 1960b. Factors necessary for *Eimeria tenella* infection of the chicken. VII. The infective forms of oocysts *(E. tenella)*. Jap. J. Vet. Sci. 22:111–122.

Itagaki, K. 1954. Further investigation on the mechanism of coccidial infection in fowl. J. Fac. Agr. Tottori Univ. 2:37–53.

Itagaki, K., and M. Tsubokura. 1958. Studies on the infectious process of coccidium in fowl. V. Further investigations on the liberation of sporozoites. Jap. J. Vet. Sci. 20: 105–110.

Jackson, A. R. B. 1962. Excystation of *Eimeria arloingi* (Marotel, 1905): stimuli from the host sheep. Nature 194:847–849.

Jackson, A. R. B. 1964. The isolation of viable coccidial sporozoites. Parasitology 54: 87–93.

Jirovec, O., K. Wenig, B. Fott, E. Bartos, J. Weiser, and R. Sramek-Husek. 1953. Protozoologie. Nak. Ceck. Akad. Ved., Prague. 643 p.

Johnson, W. T. 1933. Coccidiosis of the chicken. Oregon Agr. Exp. State Bull. 314. 16 p.

Jones, E. E. 1932. Size as a species characteristic in coccidia: variation under diverse conditions of infection. Arch. Protistenk. 76:130–170.

Joyner, L. P. 1958. Experimental *Eimeria mitis* infections in chickens. Parasitology 48: 101–112.

Kallinikova, V. D., and G. I. Roskin. 1963. Ribonucleic acid in the life cycle of *Schizotrypanum cruzi* [in Russian]. Tsitologiia 5:303–310.

Kar, A. B. 1947. Some new chemical agents for control of rabbit coccidiosis. Current Sci. 16:287–288.

Kar, A. B. 1949. In vitro action of estrogen on rabbit coccidian oocysts. Indiana Vet. J. 25:390–399.

Kheysin, Ye. M. 1935a. The structure of the oocyst and penetrability of its walls in coccidia from the rabbit [in Russian]. Tr. Inst. Pastera 2:25–45.

Kheysin, Ye. M. (Cheissin, E. M.). 1935b. Structure de l'oocyste et permeabilite de ses membranes ches les coccidies du lapin. Ann. Parasit. 13:136–146.

Kheysin, Ye. M. (Cheissin, E. M.) 1935c. Vom Einfluss anaerober Bedingungen auf verschiedene Sporulationstadien der Oocysten von *Eimeria magna* und *E. stiedae*. Arch. Protistenk. 85:426–435.

Kheysin, Ye. M. 1937a. The effects of external factors on sporulation of the oocysts of *Isospora felis* and *Isospora rivolta* [in Russian]. Tr. Inst. Pastera 3:11–20.

Kheysin, Ye. M. 1937b. Coccidia of rabbits. I. Some data on the effects of diets on infectability of rabbits by intestinal coccidia and on mortality from coccidiosis [in Russian]. Tr. Inst. Pastera 3:20–40.

Kheysin, Ye. M. 1939. Coccidia of rabbits. II. Duration of a coccidiosis invasion when rabbits are infected with the oocysts of *Eimeria magna* [in Russian]. Vestn. Mikrobiol., Epidemiol. Parazitol. 18:201–207.

Kheysin, Ye. M. 1940. Coccidiosis of rabbits. III. The developmental cycle of *Eimeria magna* [in Russian]. Uch. Zap. Len. Gos. Ped. Inst. Gertsena 30:65–91.

Kheysin, Ye. M. 1946. The duration of the life cycle of rabbit coccidia [in Russian]. Dokl. AN SSSR 52:561–564.

Kheysin, Ye. M. 1947a. Mutability of the oocysts of *Eimeria magna* [in Russian]. Zool. Zhurn. 26:17–29.

Kheysin, Ye. M. 1947b. Coccidia of the intestine of the rabbit [in Russian]. Uch. Zap. Len. Gos. Ped. Inst. Gertsena 51:1–229.

Kheysin, Ye. M. 1947c. A new species of intestinal coccidia of the rabbit—*E. coecicola* [in Russian].\Dokl. AN SSSR 55:181–183.

Kheysin, Ye. M. 1948. Development of two intestinal coccidia of the rabbit—*Eimeria piriformis* and *E. intestinalis* nom. nov. [in Russian]. m. Uch. Zap. Karelo-Finsk. Univ., III 3:179–187.

Kheysin, Ye. M. 1956. Taxonomy of the Sporozoa [in Russian]. Zool. Zhurn. 35: 1281–1298.

Kheysin, Ye. M. 1957a. Variability of the oocysts of *Eimeria intestinalis*—the parasite of the domestic rabbit [in Russian]. Vestn. LGU, ser. Biol. 2:43–52.

Kheysin, Ye. M. 1957b. Topological variations in conjugated species of coccidia of the domestic rabbit [in Russian]. Tr. Len. Obshch. Yestestvoi sp. 73:150–158.

Kheysin, Ye. M. (Cheissin, E. M.) 1958a. Cytologische Untersuchungen verschiedener Stadien des Lebenszyklus der Kaninchencoccidien. I. *Eimeria intestinalis*. Arch. Protistenk. 102:265–290.

Kheysin, Ye. M. (Cheissin, E. M.) 1958b. The role of the residual body of spores and oocysts of coccidia of the genera *Eimeria* and *Isospora*. J. Protozool. 5 (Suppl.):8.

Kheysin, Ye. M. 1959a. Observations of the residual bodies of oocysts and spores of several species of *Eimeria* from the rabbit and *Isospora* from the fox, skunk and hedgehog [in Russian]. Zool. Zhurn. 38:1776–1784.

Kheysin, Ye. M. (Cheissin, E. M.) 1959b. Cytochemical investigation of different stages of the life cycle of coccidia of the rabbit. Proc. XV Int. Congr. Zool. 713–716.

Kheysin, Ye. M. 1960. Cytological investigation of the life cycle of rabbit coccidia *Eimeria magna* [in Russian]. V. sb.: Voprosy Tsitologii i Protistologii, Izd. AN SSSR, M.-L. 258–276.

Kheysin, Ye. M. 1961. The problem of species in coccidia [in Russian]. Tr. Nauchn. Konfer. Posvyashch. 90-letiyu V. L. Yakimova. 225–230.

Kheysin, Ye. M. 1963. The Sporozoa [in Russian]. Bol'shaya Med. Ents. 31:190–203.

Kheysin, Ye. M. 1964. Electron-microscope studies of the microgametes of *Eimeria intestinalis* [in Russian]. Zool. Zhurn. 43:647–651.

Kheysin, Ye. M. (Cheissin, E. M.) 1965. Electron microscopic study of microgametogenesis in two species of coccidia from the rabbit (*Eimeria magna* and *E. intestinalis*.) Acta Protozool. 3:215–224.

Kheysin, Ye. M., and Ye. S. Snigirevskaya. (Cheissin, E. M., and E. S. Snigirevskaya) 1965. Some new data on the fine structure of the merozoites of *Eimeria intestinalis* (Sporozoa, Eimeriidea). Protisologica 1:121–126.

Kheysin, Ye. M., and V. Ye. Zaika. 1957. Some data on coccidiosis of carp [in Russian]. Tez. Soveshch. Boleznyam Ryb, Izd. AN SSSR, M.-L. 106–107.

Kheysin, Ye. M., and V. Ye. Zaika. 1960. On the specific structure of coccidia of the carp [in Russian]. Vopr. Ikhtiol. 49:193–201.

Klimeš, B. 1963. Coccidia of the domestic goose (*Anser anser dom.*). Zentralbl. Vet. Med. B. 10:427–448.

Kogan, Z. M. 1956. The influence of the soil layer on sporulation of the oocysts of chicken coccidia [in Russian]. Zool. Zhurn. 35:1454–1458.

Kogan, Z. M. 1959. Survivability of sporulated and unsporulated oocysts of chicken coccidia in winter under different conditions [in Russian]. Zool. Zhurn. 38:684–693.

Kogan, Z. M. 1960. The influence of sulfa drugs given to chickens on the ability of the oocysts of chicken coccidia to sporulate [in Russian]. Zool. Zhurn. 39:978–983.

Kogan, Z. M. 1962. Variability of shape in oocysts of chicken coccidia and its biological significance [in Russian]. Zool. Zhurn. 41:1317–1326.

Kogan, Z. M. 1965. Variability of the oocysts of chicken coccidia *Eimeria necatrix* and factors which determine it [in Russian]. Zool. Zhurn. 44:986–996.

Kotlán, S. 1932. Data on coccidiosis of water birds (duck, goose). Allat. Lapok 55:103–107.

Kotlán, S. 1932. Adatok a vízimadarak (kacsa, liba) coccidiosisának ismeretéhez. Allat. Lapok 55:103–107.

Kotlán, S., and L. Pellérdy. 1949. A survey of the species of *Eimeria* occurring in the domestic rabbit. Acta Vet. Hung. 1:93–97.

Koyama, T. 1956. Studies on the excystation of coccidial oocysts in artificial media. I. Behavior of sporozoites in the course of excystation. Dobuts. Zasshi, Tokyo 65:61–65.

Krylov, M. V. 1959a. Duration of sporogony of the oocysts of sheep coccidia on the seasonal pastures of Tadzhikistan [in Russian]. Dokl. AN TadzhSSR 2:41–43.

Krylov, M. V. 1959b. The specific structure of coccidia of domestic sheep in Tadzhikistan [in Russian]. Izv. Otd. S.-Kh. i Biol. Nauk AN TadzhSSR 1:149–175.

Krylov, M. V. 1959c. Specificity of coccidia of domestic sheep and goats [in Russian]. Mater. 10-go Soveshch. po Parazitol. Probl., Izd. AN SSSR, M.-L. 2:249–250.

Krylov, M. V 1960. Survivability of the oocysts of sheep coccidia on the seasonal pastures of Tadzhikistan [in Russian]. Izv. Otd. S.-Kh. i Biol. Nauk AN TadzhSSR 2:101–111.

Kudo, R. R. 1954. Protozoology. 4th ed. Charles C. Thomas, Springfield, Ill. 966 p.

Kupke, A. 1923. Untersuchungen über *Globidium leuckarti* Flesch. Z. Infektionskr. Haustiere 24:210–223.

Labbé, A. 1893. Sur les coccidies des oiseaux. C. R. Acad. Sci. (Paris) 117:407–409.

Lainson, R. 1959. *Atoxoplasma* Garnham, 1950, as a synonym for *Lankesterella* Labbé, 1899. Its life cycle in the English sparrow (*Passer domesticus domesticus,* Linn.) J. Protozool. 6:360–371.

Lainson, R. 1960. The transmission of *Lankesterella* ( = *Atoxoplasma)* in birds by the mite *Dermanyssus gallinae.* J. Protozool. 7:321–322.

Lainson, R. 1965. Parasitological studies in British Honduras. II. *Cyclospora niniae* sp. nov. (Eimeriidae, Cyclosporinae) from the snake *Ninia sebae sebae (Colubridae).* Ann. Trop. Med. Parasitol. 59:159–163.

Landers, E. J. 1953. The effect of low temperatures upon the viability of unsporulated oocysts of ovine coccida. J. Parsitol. 39:547–552.

Landers, E. J., Jr. 1960. Studies on excystation of coccidial oocysts. J. Parasitol. 46:195–200.

Landers, E. J., Jr. 1963. The localization of the endogenous stages of *Eimeria nieschulzi* in various host organs following the peritoneal injection of sporulated oocysts. J. Protozool. 10(Suppl.):19.

Lapage, G. 1940. The study of coccidiosis (*Eimeria caviae* [Sheather, 1924]) in the guinea pig. Vet. J. 96:144–154, 190–202, 242–254, 280–295.

Lapage, G. 1956. Veterinary parasitology. Oliver & Boyd, Ltd., Edinburgh. 964 p.

Lee, C. D. 1934. The pathology of coccidiosis in the dog. J. Amer. Med. Assoc. 85:760–781.

Lee, R. P. 1954. The occurrence of the coccidian *Eimeria bukidnonensis* Tubangui 1931, in Nigerian cattle. J. Parasitol. 40:464–466.

Léger, L., and O. Duboscq. 1910. *Selenococcidium intermedium* Lég. et Dub. et la systematique des sporozoaires. Arch. Zool. Exp. Gen. 5: 187–238.

Léger, L., and A. C. Hollande. 1912. La reproduction sexuee chez les coccidies monosporees du genre *Pfeifferinella*. Arch. Zool. Exp. Gen. 9, Not. Rev. 1:1–8.

Leuckart, K. R. 1879. Die Parasiten des Menschen und die von ihnen herrührenden Krankheiten. *In* Ein Hand-und Lehrbuch für Naturforscher und Aerzte. 2nd Ed. Leipzig and Heidelberg. 897 p.

Levine, N. D. 1957. Protozoan diseases of laboratory animals. Proc. Anim. Care Panel 7: 98–126.

Levine, N. D. 1961a. Protozoan parasites of domestic animals and of man. Burgess Publishing Co., Minneapolis. 412 p.

Levine, N. D. 1961b. Problems in the systematics of the "sporozoa." J. Protozool. 8: 442–451.

Levine, N. D. 1963. Coccidiosis. Ann. Rev. Microbiol. 17:179–198.

Levine, N. D., and V. Ivens. 1965. Species of *Isospora* in the dog. J. Parasitol. 51(Suppl.): 17.

Levine, N. D., V. Ivens, and T. E. Fritz. 1962. *Eimeria christenseni* sp. n. and other coccidia (Protozoa: Eimeriidae) of the goat. J. Parasitol. 48:255–269.

Levine, N. D., and R. N. Mohan. 1960. *Isospora* sp. (Protozoa: Eimeriidae) from cattle and its relationship to *Isospora lacazei* of the English sparrow. J. Parasitol. 46:733–741.

Levine, P. P. 1938. *Eimeria hagani* n. sp. (Protozoa: Eimeriidae) A new coccidium of the chicken. Cornell Vet. 28:263–266.

Levine, P. P. 1940. The initiation of avian coccidial infection with merozoites. J. Parasitol. 26:337–343.

Levine, P. P. 1942a. The periodicity of oocyst discharge in coccidial infection of chickens. J. Parasitol. 28:346–348.

Levine, P. P. 1942b. Excystation of coccidial oocysts of the chicken. J. Parasitol. 28: 426–428.

Levine, P. P. 1942c. A new coccidium pathogenic for chickens, *Eimeria brunetti* n. sp. Cornell Vet. 32:430–439.

Levinson, L. B., and B. T. Fedorov. 1936. Viability of the oocysts of coccidia in relation to conditions in the external environment [in Russian]. Byull. Mosk. Obshch. Ispyt. Prirody, Otd. Biol. 45:364–373.

Lickfeld, K. G. 1959. Untersuchungen über das Katzencoccid *Isospora felis* (Wenyon, 1923). Arch. Protistenk. 103:427–456.

Litver, G. M. 1935a. The retarding effects of ultraviolet rays on sporulation in coccidia [in Russian]. Vestn. Rentgenol. Radiol. 14:325–333.

Litver, G. M. (Litwer, G. M.) 1935b. Von der hemmenden Wirkung der Ultraviolett-Strahlung auf die Sporulation der Coccidien. Arch. Protistenk. 85:384–394.

Litver, G. M. (Litwer, G. M.) 1935c. Der Einfluss von geringen Dosen der Ultraviolettstrahlen auf die Stabilität des Sporulationszyklus bei Kaninchencoccidien. Arch. Protistenk. 85:395–411.

Litver, G. M. 1938a. The significance of aerobic and anaerobic conditions for the action of ultra-violet rays on the processes of sporogony in coccidia of rabbits [in Russian]. Dokl. AN SSSR 20:695–698.

Litver, G. M. 1938b. Sensitivity of the process of sporogony in coccidia oocysts to ultraviolet rays, in relation to temperature conditions, and the reparability of such oocysts [in Russian]. Dokl. AN SSSR 20:691–694.

Long, P. L. 1959. A study of *Eimeria maxima* Tyzzer, 1929, a coccidium of the fowl *(Gallus gallus)*. Ann. Trop. Med. Parasitol. 53:325–333.

Long, P. L. 1965. Development of *Eimeria tenella* in avian embryos. Nature 208:509–510.

Losanov, L. 1963. Pathomorphologische Untersuchungen bei Spontaner kokzidiose der Kucken. C. R. Acad. Bulg. Sci. 16:549–552.

Löser, E. and R. Gönnert. 1965. Zur Bildung der Sklerotinhülle der Oocysten einiger Coccidien. Z. Parasitenk. 25:597–605.

Lotze, J. C. 1953. Life history of the coccidian parasite, *Eimeria arloingi,* in domestic sheep. Amer. J. Vet. Res. 14:86–95.

Lotze, J. C., and R. G. Leek. 1960. Some factors involved in excystation of the sporozoites of three species of sheep coccidia. J. Parasitol. 46(Suppl.):46–47.

Lotze, J. C., and R. G. Leek. 1963. Excystation of coccidial parasites in various animals. J. Parasitol. 49(Suppl.):32.

Lotze, J. C., W. T. Shalkop, R. G. Leek, and R. Behin. 1964. Coccidia schizonts in mesenteric lymph nodes of sheep and goats. J. Parasitol. 50:205–208.

Ludvik, J. 1958. Morphology of *Toxoplasm gondii* in electron microscope. Vestn. Ceskosl. Zool. Spol. 22:130–136.

Ludvik, J. 1963. Electron microscopic study of some parasitic Protozoa. 1st Int. Congr. Protozool. Prague, 1961, Proc. 387–393.

Machinsky, A. P. 1954. A development of several questions of epizootiology, treatment and prophylaxis of chickens infected with coccidia [in Russian]. Avtoref. Kand. Diss., M.

Marinček, M. 1965. *Eimeria subepithelialis* chez la carpe. *In* Progress in Protozoology. 2nd Int. Conf. Protozool., London, Abstr. 160–162.

Marotel, G. 1906. La coccidiose de la chèvre et son parasite. Ann. Méd. Vét. 55:171. (Abstr.)

Marotel, G. 1949. Parasitologie vétérinaire. Parasites et maladies parasitaires des animaux. Bailliere et Fils, Paris. 652 p.

Marquardt, W. C. 1957. The effect of temperature on the sporulation of *Eimeria zurnii* of cattle. J. Protozool. 4(Suppl.):10.

Marquardt, W. C. 1960a. Oocyst production in subclinical infections with coccidia in cattle. J. Parasitol. 46(Suppl.):46.

Marquardt, W. C. 1960b. Effect of high temperature on sporulation of *Eimeria zurnii.* Exp. Parasitol. 10:58–65.

Marquardt, W. C. 1963. Observations on living *Eimeria nieschulzi* of the rat. J. Parasitol. 49(Suppl.):28.

Marquardt, W. C., and M. M. Kallor. 1962. The effect of temperature and oxygen on the development of oocysts of *Eimeria nieschulzi.* J. Protozool. 9(Suppl.):9.

Marquardt, W. C., C. M. Senger, and L. Seghetti. 1960. The effect of physical and chemical agents on the oocysts of *Eimeria zurnii* (Protozoa, Coccidia). J. Protozool. 7: 186–189.

Matsubayashi, H. 1934. Studies of the life history and classification of *Eimeria* of the rabbit. Keio-Igaku, Tokyo. 14:513–560.

Matsubayashi, H. 1935. *Eimeria* parasitic in marine fishes. Keio-Igaku, Tokyo. 15: 1281–1300.

Melikyan, Ye. D. 1954. Duration of survival of coccidia in the external environment [in Russian]. Veterinariya 5:43–44.

Metelkin, A. I. 1936. Modern achievements in the study of rabbit coccidia [in Russian]. Zhurn. Epidemiol. Immunol. 16:309–320.

Metzner, R. 1903. Untersuchungen an *Coccidium cuniculi.* Arch. Protistenk. 2:13–72.

Monné, L., and G. Hönig. 1954. On the properties of the shells of the coccidian oocysts. Ark. Zool. 7:251–256.

Moore, E. N. 1954. Species of coccidia affecting turkeys. 91st Annu. Meet. Amer. Vet. Med. Assoc., Proc. 300–304.

Moore, E. N., and J. A. Brown. 1951. A new coccidium pathogenic for turkeys, *Eimeria adenoeides* n. sp. (Protozoa: Eimeriidae). Cornell Vet. 41:124–135.

Moore, E. N., and J. A. Brown. 1952. A new coccidium of turkeys, *Eimeria innocua* n. sp. (Protozoa: Eimeriidae). Cornell Vet. 42:395–402.

Moore, E. N., J. A. Brown, and R. D. Carter. 1954. A new coccidium of turkeys, *Eimeria subrotunda* n. sp. (Protozoa: Eimeriidae). Poultry Sci. 33:925–929.

Morgan, B. B., and P. A. Hawkins. 1948. Veterinary protozoology. Burgess Publishing Co., Minneapolis. 195 pp.

Mosevich, T. N., and Ye. M. Kheysin. 1961. Some data on electron microscopy of the merozoites of *Eimeria intestinalis* [in Russian]. Tsitologiia 3:34–39.

Moulder, J. W. 1962. The biochemistry of intracellular parasitism. University of Chicago Press, Chicago. 171 p.

Mukherjea, A. K., and H. N. Ray. 1962. A method for staining oocysts of a coccidium *Eimeria tenella* from chicks. Bull. Calcutta Sch. Trop. Med. 10:78–79.

Musayev, M. A., and A. M. Veysov. 1965. Coccidia of rodents of the USSR [in Russian]. Baku. 1–152.

Nabih, A. 1938. Studien über die Gattung Klossia und Beschreibung des Lebenszyklus von *Klossia loossi* (nov. sp.) Arch. Protistenk. 91:474–515.

Nath, V., G. P. Dutta, and O. Sagar. 1960. The life-cycle of *Eimeria tenella* Railliet and Lucet, 1891, and its variations during the experimental infection of chicks. Res. Bull. (N.S.), Panjab Univ. 11:227–235.

Nath, V., G. P. Dutta, and O. Sagar. 1962. Observations on the macrogametocyte leading to the formation of the oocyst in *Eimeria tenella,* Railliet and Lucet, 1891, in experimentally infected chicks. Res. Bull. (N.S.). Panjab Univ. 12:215–220.

Nath, V., G. P. Dutta, and O. Sagar. 1965. Phase-contrast microscope studies on *Eimeria tenella.* Res. Bull. Panjab Univ. 16:151–157.

Naville, A. 1925. Recherches sur le cycle sporogonique des *Aggregata.* Rev. Suisse Zool. 32:125–179.

Naville, A. 1927. Recherches sur le cycle évolutif et chromosomique de *Klossia helicina* (A. Schneider). Arch. Protistenk. 57:427–474.

Neméséri, L. 1960. Beiträge zur Ätiologie der Coccidiose der Hunde. I. *Isospora canis* sp. n. Acta Vet. Acad. Sci. Hung. 10:95–99.

Nieschulz, O. 1922. Über die Benennung des Schweinecoccids. Centr. Bakt. I. Abt. Orig. 88:379–380.

Nöller, W. 1923. Zur Kenntnis eines Nierencoccids. Der Entwicklungskreis des Coccids der Wasserfroschniere. (*Isospora lieberkuhni* [Labbé, 1894]). Arch. Protistenk. 47: 101–108.

Nyberg, P. A., and D. M. Hammond. 1964. Excystation of *Eimeria bovis* and other species of bovine coccidia. J. Protozool. 11:474–480.

Ohkubo, Y., M. Ikeda, and K. Tsunoda. 1955. Coccidiose des poussins. Bull. Office Intern. Epizoot. 44:216–231.

Orlov, N. P. 1956. Coccidia of Farm Animals [in Russian]. Sel'khozgiz, M. 1–165.

Orlov, N. P., and P. U. Aytykina. 1936. The problem of chemical disinfection in coccidiosis of rabbits [in Russian]. Tr. Alma-Atinsk. Zoovet. Inst. 2:65–71.

Pande, B. P., B. B. Bhatia, and K. M. N. Srivastava. 1965. *Wenyonella anatis* n. sp. from Indian domestic duck. Sci. Culture 31:383–384.

Patterson, F. D. 1933. Studies on the viability of *Eimeria tenella* in soil. Cornell Vet. 23: 232–249.

Pattillo, W. H. 1959. Invasion of the cecal mucosa of the chicken by sporozoites of *Eimeria tenella*. J. Parasitol. 45:253–258.

Pattillo, W. H., and E. R. Becker. 1955. Cytochemistry of *Eimeria brunetti* and *E. acervulina* of the chicken. J. Morphol. 96:61–96.

Patton, W. H. 1965. *Eimeria tenella:* cultivation of the asexual stages in cultured animal cells. Science 150:767–769.

Pellérdy, L. 1949. Studies on coccidia occurring in the domestic pig with the description of a new *Eimeria* species (*E. polita* sp. n.) of that host. Acta Vet. Acad. Sci. Hung. 1: 101–109.

Pellérdy, L. 1953. Beiträge zur Kenntnis der Darmkokzidiose des Kaninchens. Die endogene Entwickelung von *Eimeria piriformis*. Acta Vet. Acad. Sci. Hung. 3:365–376.

Pellérdy, L. 1954a. Contribution to the knowledge of coccidia of the common squirrel *(Sciurus vulgaris)*. Acta Vet. Acad. Sci. Hung. 4:475–480.

Pellérdy, L. 1954b. Beitäge zur Spezifität der Coccidien des Hasen und Kaninchens. Acta Vet., Budapest 4:481–487.

Pellérdy, L. 1956. On the status of the *Eimeria* species of *Lepus europaeus* and related species. Acta Vet. Acad. Sci. Hung. 6:451–467.

Pellérdy, L. 1960. Intestinal coccidiosis of the coypu. II. The endogenous development of *Eimeria seideli* and the present status of the group *"Globidium"*. Acta Vet. Acad. Sci. Hung. 10:389–399.

Pellérdy, L. 1965. Coccidia and coccidiosis. Académiai Kiadó, Budapest. 657 p.

Pellérdy, L., and A. Babos. 1953. Untersuchungen über die endogene Entwicklung sowie pathologische Bedeutung von *Eimeria media*. Acta Vet. Acad. Sci. Hung. 3:173–188.

Pérard, C. 1924. Recherches sur les coccidies et les coccidioses du lapin. C. R. Acad. Sci. (Paris) 178:2131–2134.

Pérard, C. 1925. Recherches sur les coccidies et les coccidioses du lapin. II. Contribution a l'etude de la biologie des oocystes de coccidies. Ann. Inst. Pasteur, Paris 39:505–542.

Peraza, L. A., de. 1963. Studies on two new coccidia from the Venezuelan lizard *Cnemidophorus lemniscatus lemniscatus*. *Hoarella garnhami* gen. nov., sp. nov. and *Eimeria flaviviridis americana* subsp. nov. Parasitology 53:95–107.

Petrov, V. A. and N. N. Nikonov. 1964. Coccidiosis of cattle [in Russian]. Izd. "Kolos," M. 1–70.

Pratt, I. 1937. Excystation of the coccidia, *Eimeria tenella*. J. Parasitol. 23: 426–427.

Ray, H. N. 1930. Studies on some Sporozoa in polychaete worms. II. *Dorisiella scolelepidis*, n. gen. n. sp. Parasitology 22:471–480.

Ray, H. N. 1945. On a new coccidium *Wenyonella gallinae* n. sp. from the gut of the domestic fowl, *Gallus gallus domesticus* Linn. Curr. Sci. 14:275.

Ray, H. N. and B. S. Gill. 1954. Prelininary observations on alkaline phosphatase in experimental *Eimeria tenella* infection in chicks. Ann. Trop. Med. Parasitol. 48:8–10.

Ray, H. N. and B. S. Gill. 1955. Observations on the nucleic acids of *Eimeria tenella* Railliet and Lucet, 1891. Indian J. Vet. Sci. Anim. Husb. 25:17–23.

Reich, F. 1913. Das Kaninchencoccid *Eimeria stiedae* nebst einem Beitrage zur Kenntnis von *Eimeria falciformis* (Eimer, 1870). Arch. Protistenk. 28:1–42.

Reichenow, E. 1921. Die Hämococcidien der Eidechsen. Vorbemerkungen und I. Teil: Die Entwicklungsgeschichte von *Karyolysus*. Arch. Protistenk. 42:179–291.

Reichenow, E. 1931. Die Coccidien. *In* Handbuch der pathogenen Protozoen. Johann Ambrosius Barth, Leipzig. Prowazek, S. von, Ed. p. 1136–1277.

Reichenow, E. 1932. Sporozoa *In* Die Tierwelt der Nord-und Ostsee. Grimpe, G. and E. Wagler, Eds., Vol. 21. Academische Verlagsgesellschaft, Leipzig.

Reinhardt, J. F., and E. R. Becker. 1933. Time of exposure and temperature as lethal factors in the death of the oocysts of *Eimeria miyairii*, a coccidium of the rat. Iowa State Coll. J. Sci. 7:505–510.

Reyer, W. 1937. Infektionsversuche mit *Barrouxia schneideri* an *Lithobius forficatus*, insbesondere zur Frage der sexualität der Coccidiensporozoiten. Zeitschr. Protistenk. 9: 478–522.

Richardson, U. F., and S. B. Kendall. 1957. Veterinary protozoology. Oliver & Boyd, Ltd., Edinburgh. 260 p.

Rootes, D. G. and P. L. Long. 1965. Studies on the synthesis of glycogen by *Eimeria necatrix* in domestic fowl. *In* Progress in protozoology. 2nd Int. Conf. Protozool., London, Abstr. 157–158.

Roskin, G. I., and L. V. Balicheva. 1961. Cytochemistry of nucleotids and nucleic acids in the cells of the liver of the axolotl [in Russian]. Tsitologiia 3:305–311.

Roudabush, R. L. 1935. Merozoite infection in coccidiosis. J. Parasitol. 21: 453–454.

Roudabush, R. L. 1937. The endogenous phases of the life cycles of *Eimeria nieschulzi*, *Eimeria separata*, and *Eimeria miyairii*, coccidian parasites of the rat. Iowa State Coll. J. Sci. 11:135–163.

Rudzinska, M. A., and W. Trager. 1957. Intracellular phagotrophy by malaria parasites: an electron microscope study of *Plasmodium lophurae*. J. Protozool. 4:190–199.

Rutherford, K. L. 1943. The life cycle of four intestinal coccidia of the domestic rabbit. J. Parasitol. 29:10–32.

Schaudinn, F. 1900. Untersuchungen über den Generationswechsel bei Coccidien. Zool. Jahrb., Abt. Anat. 13:197–292.

Schaudinn, F. 1902. Studien über krankheitserregende Protozoen. I. *Cyclospora caryolytica* Schaud., der Erreger der perniciösen enteritis des Maulwurfs. Arb. Kais. Gesundh. 18: 378–416.

Schaudinn, F. 1905. Neuere Forschungen über die Befruchtung bei Protozoen. Verhandl. Deutsch. Zool. Gesellsch. 15 Jahresversamml., Breslau, 13–16 June, 16–35.

Scholtyseck, E. 1953. Beitrag zur kenntnis des Entwicklungsganges des Hühnercoccids *Eimeria tenella*. Arch. Protistenk. 98:415–465.

Scholtyseck, E. 1954. Untersuchungen über die bei einheimischen Vogelarten vorkommenden Coccidien der Gattung *Isospora*. Arch. Protistenk. 100:91–112.

Scholtyseck, E. 1955. *Eimeria anatis* n. sp. ein neues Coccid aus der Stockente (*Anas platyrhynchos*). Arch. Protistenk. 100:431–434.

Scholtyseck, E. 1962a. Über die Feinstruktur von *Eimeria perforans* (Sporozoa). Z. Parasitenk. 22:123–132.

Scholtyseck, E. 1962b. Electron microscope studies on *Eimeria perforans* (Sporozoa). J. Protozool. 9:407–414.

Scholtyseck, E. 1963a. Vergleichende Untersuchungen über die Kernverhältnisse und das Wachstum bei Coccidiomorphen unter Besonderer Berücksichtigung von *Eimeria maxima*. Z. Parasitenk. 22:428–474.

Scholtyseck, E. 1963b. Elektronmikroskopische Untersuchungen über die Wechselwirkung zwischen dem Zellparasiten *Eimeria perforans* und seiner Wirtzelle. Z. Zellforsch. 61: 220–230.

Scholtyseck, E. 1964. Elektronmikroskopisch-cytochemischer Nachweis von Glykogen bei

*Eimeria perforans.* Z. Zellforsch. 64:688–707.

Scholtyseck, E. 1965a. Die Mikrogametenentwicklung von *Eimeria perforans.* Z. Zellforsch. 66:625–642.

Scholtyseck, E. 1965b. Elektronenmikroskopische Untersuchungen über die Schizogonie bei Coccidien (*Eimeria perforans* und *E. stiedae*) Z. Parasitenk. 26:50–62.

Scholtyseck, E., and G. Piekarski. 1965. Elektronenmikroskopische untersuchungen an Merozoiten von Eimerien (*Eimeria perforans* und *E. stiedae*) und *Toxoplasma gondii.* Zur systematischen Stellung von *T. gondii.* Z. Parasitenk. 26:91–115.

Scholtyseck, E., and D. Schäfer. 1963. Über schlauchförmige ausstülpungen an der Zellmembran der Makrogametocyten von *Eimeria perforans.* Z. Zellforsch. 61:214–219.

Scholtyseck, E., and D. Speicker. 1964. Vergleichende elektronenmikroskopsiche Untersuchungen an den Entwicklungsstadien von *Eimeria perforans* (Sporozoa). Z. Parasitenk. 24:546–560.

Scholtyseck, E., and W.-H. Voigt. 1964. Die Bildung der Oocystenhülle bei *Eimeria perforans* (Sporozoa). Z. Zellforsch. 62:279–292.

Scholtyseck, E., and N. Weissenfels. 1956. Elektronenmikroskopische Untersuchungen von Sporozoen. I. Die Oocystenmembran des Huhnercoccids *Eimeria tenella.* Arch. Protistenk. 101:215–222.

Schwalbach, G. 1959. Untersuchungen und Beobachtungen an Coccidien der Gattungen *Eimeria, Isospora* und *Caryospora* bei Vögeln mit einer Beschreibung von sechzehn neuen Arten. Arch. Protistenk. 104:431–491.

Schwalbach, G. 1961a. Die Coccidiose der Singvögel. II. Beobachtungen an *Isospora*-oocysten aus einem Weichfresser (*Parus major*) mit besonderer Berücksichtigung des Ausscheidungsrhythmus. Zentralbl. Bakteriol. 181:264–279.

Schwalbach, G. 1961b. Die Coccidiose der Singvögel. III. Die Temperaturabhängigkeit der exogenen Entwicklungsphase. Zentralbl. Bakteriol. 183:272–282.

Senger, C. M. 1959. Chemical inhibition of sporulation of *Eimeria bovis* oocysts. Exp. Parasitol. 8:244–248.

Sharma, N. N., and W. M. Reid. 1962. Successful infection of chickens after parenteral inoculation of oocysts of *Eimeria* sp. J. Parasitol. 48(Suppl.):33.

Sheffield, H. G., and D. M. Hammond. 1965. The fine structure of first generation merozoites of *Eimeria bovis. In* Progress in Protozoology. 2nd Int. Conf. Protozool., London, Abstr. 157.

Shevchenko, M. Ye. 1953. The dynamics of eimeriosis of sheep in the Chkalovsk Oblast' [in Russian]. Tr. Chkalovsk. S.-Kh. Inst. 6:155–158.

Shiyanov, A. T. 1954. On specificity in coccidia [in Russian]. Tr. Przheval'sk. Gos. Ped. Inst. 3:185–190.

Shul'man, S. S., and V. Ye. Zaika. 1962. Coccidia. *In* A guide to the parasites of freshwater fishes in the USSR [in Russian]. Red. B. Ye. Bykhovskoy. Izd. AN SSSR, M.-L: 29–46.

Shumard, R. F. 1957. Ovine coccidiosis—incidence, possible endotoxin, and treatment. J. Amer. Vet. Med. Assoc. 131:559–561.

Sibalić, S. 1949. La development des sporozoites dans les oocystes chez *Eimeria stiedae* et *E. perforans.* Arkh. Biol. Nauka, Beograd 1:253–257.

Skidmore, L. V. 1929. Note on a new species of coccidia from the pocket gopher (*Geomys bursarius*) (Shaw). J. Parasitol. 15:183–184.

Skidmore, L. V. 1933. Bovine coccidia carriers. J. Parasitol. 20:126.

Smetana, H. 1933a. Coccidiosis of the liver in rabbits. I. Experimental study on the excystation of oocysts of *Eimeria stiedae.* Arch. Pathol. 15:175–192.

Smetana, H. 1933b. Coccidiosis of the liver in rabbits. II. Experimental study on the mode of infection of the liver by sporozoites of *Eimeria stiedae*. Arch. Pathol. 15:330–339.

Smetana, H. 1933c. Coccidiosis of the liver in rabbits. III. Experimental study of the histogenesis of coccidiosis of the liver. Arch. Pathol. 15:516–636.

Sneed, K. and G. Jones. 1950. A preliminary study of coccidiosis in Oklahoma quail. J. Wildl. Manage. 14:169–174.

Steinert, M., and A. W. Novikoff. 1960. The existence of a cytostome and the occurrence of pinocytosis in the trypanosome, *(Trypanosoma mega)*. J. Biophys. Biochem. Cytol. 8: 563–570.

Strout R. G., H. Botero, S. C. Smith, and W. R. Dunlop. 1963. The lipids of coccidial oocysts. J. Parasitol. 49(Suppl.):20.

Strout, R. G., J. Solis, S. C. Smith, and W. R. Dunlop. 1965. In vitro cultivation of *Eimeria acervulina* (Coccidia). Exp. Parasitol. 17:241–246.

Svanbayev, S. K. 1955. A new species of coccidia of turkeys [in Russian]. Tr. AN KazSSR 3:161–163.

Tanabe, M. 1938. On three species of coccidia of the mole, *Mogera wogura coreana* Thomas, with special reference to the life history of *Cyclospora caryolytica*. Keijo J. Med. 9–21:52.

Tomimura, T. 1957. Experimental studies on coccidiosis in dogs and cats. I. The morphology of oocyst and sporogony of *Isospora felis*, and its artificial infection in cats. Kiseichugaku Zasshi 6:12–24.

Trigg, P. I. 1965. The life cycle and pathogenicity of *Eimeria phasiani* from the pheasant. Progress in Protozoology. 2nd Int. Conf. Protozool. London, Abstr. 159–160.

Tsunoda, K. 1952. *Eimeria matsubayashii* sp. nov. a new species of rabbit coccidium. Exp. Rep. Gov. Exp. Sta. Anim. Hyg. 25:109–119.

Tsunoda, K., and O. Itikawa. 1955. Histochemical studies of chicken coccidia *(Eimeria tenella)*. I. On the nucleic acids, polysaccharides and phosphomonoesterases in their several developmental stages. Exp. Rev. Gov. Exp. Sta. Anim. Hyg. 29:73–82.

Tsygankov, A. A. 1950. Data on a study of coccidia of the camel [in Russian]. Izv. AN KazSSR, Ser. Parazitol. 8:174–180.

Tsygankov, A. A., N. G. Paychuk, and Z. A. Balbayeva. 1959. Materials towards the question of the specificity of coccidia of sheep, goats, and the saiga antelope [in Russian]. Mater. 10-go Soveshch. po Parazitol. Probl. Izd. AN SSSR, M.-L., 2:265.

Tuzet, O. and C. Bessiere. 1945. Sur la *Legerella testiculi* Cuenot, Coccidie parasite du testicule du *Glomeris marginata* Villers. Arch. Zool. Exper. Gen. 84, N. R.: 70–77.

Tyzzer, E. E. 1910. An extracellular coccidium, *Cryptosporidium muris* (gen. et sp. nov.), of the gastric glands of the common mouse. J. Med. Res. 23:487–509.

Tyzzer, E. E. 1912. *Cryptosporidiim parvum* (sp. nov.), a coccidium found in the small intestine of the common mouse. Arch. Protistenk. 26:394–412.

Tyzzer, E. E. 1927. Species and strains of coccidia in poultry. J. Parasitol. 13:215.

Tyzzer, E. E. 1929. Coccidiosis in gallinaceous birds. Amer. J. Hyg. 10:269–383.

Tyzzer, E. E., H. Theiler, and E. E. Jones. 1932. Coccidiosis of gallinaceous birds. II. A comparative study of species of *Eimeria* of the chicken. Amer. J. Hyg. 15:319–393.

Ugolev, A. M. 1963. Pristencochnoye (kontaktnoye) pishchevareniye. (Membranic [contact] digestion.) Izd. Nauka, M.-L.:1–170.

Uricchio, W. A. 1953a. The feeding of artificially altered oocysts of *Eimeria tenella* as a means of establishing immunity to cecal coccidiosis in chickens. Proc. Helm. Soc. Wash. 20:77–83.

Uricchio, W. A. 1953b. Destruction of coccidial oocysts of chickens by means of chemicals. Exp. Parasitol. 2:16–18.

Vagin, A. N. 1930. Coccidiosis of rabbits [in Russian]. Krolikovodstvo, 10.

Van Doorninck, W. M., and E. R. Becker. 1957. Transport of sporozoites of *Eimeria necatrix* in macrophages. J. Parasitol. 43:40–44.

Vetterling, J. M. 1965. The coccidia (Protozoa: Eimeriidae) of swine. Diss. Abstr. 26: 564–565.

Vetterling, J. M. 1966. Endogenous cycle of the swine coccidium *Eimeria debliecki* Douwes, 1921. J. Protozool. 13:290–300.

Vil'son, E. 1936. The cell and its role in development and inheritance [in Russian]. t. I. Izd. Biol. Med. Liter., M.-L.:1–564.

Vivier, E. 1963. Une nouvelle coccidie, *Eucoccidium durchoni* n. sp. parasite de l'annelide *Nereis diversicolor*. 1st Int. Congr. Protozool., Prague, 1961, Proc. 449–451.

Vivier, E., and E. Henneré. 1964a. Cytologie, cycle et affinités de la coccidie *Coelotropha durchoni*, nomen novum ( = *Eucoccidium durchoni* Vivier), parasite de *Nereis diversicolor* O. F. Müller (Annelide Polychete). Bull. Biol. France Belgique 98:153–206.

Vivier, E., and E. Henneré. 1964b. Observation sur le cycle de certaines coccidis parasites de polychetes. Arch. Zool. Exp. Gen. 104:194.

Vivier, E., and E. Henneré. 1965. Ultrastructure des stades végétatifs de la coccidie *Coelotropha durchoni*. Protistologica 1:89–105.

Wagner, W. H. 1965. Comparative investigations on morphology and tissue-migration of *E. tenella* and *E. acervulina* by means of PAS-AO-method. *In* Progress in protozoology. 2nd Int. Conf. Protozool., London, Abstr. 156–157.

Walton, A. C. 1959. Some parasites and their chromosomes. J. Parasitol. 45:1–20.

Wasielewski, T. 1904. Studien und Mikrophotogramme zur Kenntnis der Pathogenen Protozoen., 1 Heft. 175 p.

Waworuntu, F. K. 1924. Bijdrage tot de kennis van het konijnencoccidium. Diss. Utrecht: 1–95.

Waxler, S. H. 1941. Immunization against cecal coccidiosis in chickens by the use of X-ray-attenuated oocysts. J. Amer. Vet. Med. Assoc. 99:481–485.

Wedekind, G. 1927. Zytologische Untersuchungen an *Barrouxia schneideri* (Gametenbildung, Befruchtung und Sporogonie), zugleich ein Beitrag zum Reduktionsproblem. (Coccidienuntersuchungen I). Z. Zellforsch. Mikr. Anat. 5:505–595.

Wenyon, C. M. 1926. Protozoology. London. Vol. 1 and 2. 1563 p.

Wenyon, C. M., and L. Scheather. 1925. Exhibition of specimens illustrating *Isospora* infections of dogs. Roy. Soc. Trop. Med. Hyg., Trans. 19:10.

Wiesenhütter, E. 1962. Eim beitrag zur Kenntnis der endogenen Entwicklung von *Eimeria spinosa* des Schweines. Berlin Munchen Tieraerztl. Wochenschr. 75:172–173.

Wilson, I. D. 1931. A study of bovine coccidiosis. Va. Agr. Exp. Sta., Techn. Bull. 42.

Wilson, I. D., and L. C. Morley. 1933. A study of bovine coccidiosis, II. J. Amer. Vet. Med. Assoc. 35:826–850.

Wilson, P. A. G., and D. Fairbairn. 1961. Biochemistry of sporulation in oocysts of *Eimeria acervulina*. J. Protozool. 8:410–416.

Yakimov, V.L. (Yakimoff, W. L.). 1929. Zur Frage über den Parasitismus der Süsswasserfische. IV. Coccidien beim Barsch *(Perca fluviatilis)*. Arch. Protistenk. 67:501–508.

Yakimov, V. L. 1931. Veterinary Protozoology [in Russian]. Sel'khozgiz, M.-L.: 1–863.

Yakimov, V. L., and I. G. Galuzo. (Yakimoff, W. L. and J. G. Galouzo). 1927. Zur Frage über Rindercoccidien. Arch. Protistenk. 58:185–200.

Yakimov, V. L. and P. Timofeyev. 1940. On the inclusions found in the oocysts of *Eimeria labbeana* [in Russian]. Vestn. Mikrobiol. Epidemiol. Parazitol. 19:150–152.

Yarwood, E. A. 1937. The life cycle of *Adelina cryptocerci* sp. nov., a coccidian parasite of the roach *Cryptocercus punctulatus*. Parasitology 29:370–390.

Young, B. P. 1929. A quantitative study of poultry coccidiosis, with data on the prepatent and patent periods in the life cycle of *Eimeria avium*. J. Parasitol. 15:241–250.

Zaika, V. Ye., and Ye. M. Kheysin. 1959. Coccidiosis of carp at the Valday Fish Farm [in Russian]. Izv. Vsesoyuzn. N. -Issl. Inst. Ozern. Rechn. Rybn. Khoz. 49:221–224.

Zasukhin, D. N. (Sassuchin, D. N.) 1935. Zum Studium der Protisten-und Bacterienkerne. I. Über die Nucleareaction und ihre Anwendung bei protozoologischen und bacteriologischen Untersuchungen. Arch. Protistenk. 84:186–198.

Zellen, B. W., von. 1959. *Eimeria carolinensis* n. sp., a coccidium from the white-footed mouse, *Peromyscus leucopus* (Rafinesque). J. Protozool. 6:104–105.

Zmerzlaya, Ye. I. 1965. Coccidiosis enteritis in carp [in Russian]. Avtoref. Diss. L.: 1–16.

# Index